T0091928

Introduction to Computational Cardiology

Boris Ja. Kogan

Introduction to Computational Cardiology

Mathematical Modeling and Computer Simulation

Boris Ja. Kogan
Department of Computer Science
University of California, Los Angeles
4731 Boelter Hall
Los Angeles, CA 90095-1596
USA
kogan@cs.ucla.edu

ISBN 978-0-387-76685-0 e-ISBN 978-0-387-76686-7
DOI 10.1007/978-0-387-76686-7
Springer New York Dordrecht Heidelberg London

Library of Congress Control Number: 2009942421

Printed on acid-free paper

Springer is part of Springer Science+Business Media (www.springer.com)

I dedicate this book to the memory of my parents Jacob and Rebekka Kogan, who gave me life twice.

I also owe a tribute to my wife Mina Rajskina-Kogan, Professor of cardiology, MD, PhD. She drew my attention and interest to the subjects treated in this book and patiently waited for the accomplishment of the work.

Preface

The main objective of this text is to provide a comprehensive, in-depth introduction to the background materials, fundamental concepts, and research challenges in mathematical modeling and computer simulation of electrophysiological heart processes, under both normal and pathological conditions.

Though these topics are closely connected to the field of scientific computing they are interdisciplinary in nature, combining the disciplines of applied mathematics, computer science, physiology, and medicine. From a theoretical point of view these topics are related to the study of a particular type of the more general class of so-called excitable media.

An excitable medium is defined as a nonlinear dynamical system consisting of units distributed in space that interact according to the laws of diffusion. Each unit of the system is itself a nonlinear dynamical system, with its own source of energy. After application of an external over-threshold stimulus, a unit can generate a solitary excitation pulse or a sequence of pulses depending on its nonlinear properties.

The following specific wave phenomena can be observed in excitable media:

- Propagation of traveling waves without decay
- The formation of spiral waves
- Generation of circular waves by autonomous leading centers
- Formation of dissipative structures (e.g. stationary standing wave)

Excitable media are encountered in both biology (cardiac tissue, nerve fiber, smooth and skeletal muscles, eye retina, population of amoeboid cells, some demographical problems, etc.) and technology (specific chemical reactions, some microelectronics devices, burning processes, plasma systems, etc.).

Mathematically, the processes in excitable media are described by a special type of parabolic nonlinear partial differential equations known in literature as reaction-diffusion equations. In contrast to engineering systems, in which mathematical models (descriptions) can be derived from the first principles, biological systems are described in a majority of cases as semi-phenomenological mathematical models partly derived from the first principles and partly represented as the mathematical approximation of experimental data obtained for a specific situation and extrapolated to the more general ones.

The scope of this text is restricted to the mathematical modeling and computer simulations of the dynamical processes in a particular class of these systems – cardiac cells and tissue. It is divided on the following major parts:

I. Electrophysiological background and basic concepts of mathematical modeling and computer simulation.

II. Mathematical modeling and computer simulation of action potential (AP) generation, from simple models such as Van der Pol and FitzHugh-Nagumo to physiological models of the Ist and IInd generations based on the Hodgkin-Huxley formalism.

III. Theory, mathematical modeling, and computer simulations of excitation wave propagation in one-dimensional tissue

IV. Mathematical modeling and computer simulations of excitation wave propagation in uniform and non-uniform two-dimensional tissues including rectilinear and circular wave propagation, theory of stationary and nonstationary spiral waves, and conditions of wave front breakup as analogue of tachyarrhythmia and ventricular fibrillation.

V. The implementation of mathematical models on serial and parallel supercomputers.

Special attention is devoted to new topics such as Markovian representation of cell channel gate processes, and new phenomena appearing in single cells with Ca dynamics under high pacing rates and in cardiac tissues during spiral wave propagation.

The included topics do not cover such important subjects as propagation in three-dimensional tissue with natural heterogeneity of AP characteristics along the thickness of the tissue and directional variability of fiber angles. Computer simulations of these problems have until now been performed using over-simplified non-physiological models such as FitzHugh-Nagumo and simplified ionic models such as the Luo-Rudy I AP model. Application of more realistic models is under intensive investigation. A similar situation is encountered in the investigation of the effect of mechanical cell contractions on conductivity of cell channels and AP propagation. The dynamics of the pacemaker system and the development of cell contraction processes, described in detail in published books, are excluded from the text and replaced by corresponding references.

Despite the many talented scientists working in mathematical modeling and computer simulation of cardiac processes, there are currently no published materials in which these topics are treated systematically, up-to-date with current research and containing the required minimum of materials which allow the specialists in other fields (mathematics, computer science, heart physiology and cardiology) to participate in such interdisciplinary research.

The book, "Simulation of Wave Processes in Excitable Media", by Dr. Zykov (my former PhD student and later colleague) was published in 1984 in the Soviet Union. The English translation of this book (from Russian) was edited by late Dr. Winfree and published in 1987. It includes some of the first approaches and information on the subject matter. The content of the book has become obsolete (except for the kinematics theory of stationary spiral waves), especially as applied to heart processes, because it focuses on simplified, nonionic cell models in which Ca dynamics are not present. J. Keener and J. Sneid's excellent work, "Mathematical Physiology," is devoted to broad topics of mathematical modeling of different physiological systems. Unfortunately the authors, in taking a more general approach, did not consider the heart processes in detail and fully omitted the implementation of mathematical models on modern parallel computer systems, focusing instead on the use of standard programs on desktop computers. Several collections of papers exist (e.g. "Computational Biology of the Heart," edited by A.V. Panfilov and A.V. Holden, 1997) addressing some of the proposed topics, but these collections require extensive prior knowledge of the subject and as such are not functional as an introductory text.

The content of this manuscript is combined from the author's lecture notes for the course "Introduction to Computational Cardiology," delivered to graduate students of the UCLA Computer Science and Biomedical Engineering Departments; the results of his personal research activities and those conducted by his PhD students in the former Soviet Union and United States over the last 30-35 years; and new achievements described in current literature. This book can serve not only as a text book for graduate students specializing in modeling and computer simulation of heart processes, but also as a reference for researchers engaged in mathematical modeling and computer simulation of different bio-medical problems. The latter, among other useful information, may find in the text many challenging problems awaiting solutions.

Acknowledgements

I would like to express my great thanks to my colleagues and friends Drs. W. Karplus, G. Estrin, L. Kleinrock, M. Melkanoff, M. Erzegovac (UCLA Computer Science Department) and Drs. G. Bekey (USC) and G. Korn (University of Arizona), for their invaluable help in my resettling in the United States, helping me to continue my scientific career and, by extension, the writing of this text.

For the last twenty years, I have been a part of UCLA and wish to thank all faculty members and staff of the Computer Science Department for their unwavering support.

The collaboration with the cardiovascular laboratory in the Department of Cardiology at UCLA and personally with Drs. J. Weiss, J. Goldhaber, H. Karagueuziand, A. Garfinkel and S. Lamp over the last 15 years gave me additional knowledge of heart physiology and a new understanding of cardiac problems. They also provided me with constructive critiques. Participation in two long term NIH grants (PI, Dr. Weiss) allowed me to support the research activity of some of my PhD students.

All studies of wave propagation along 2- and 3-dimentional cardiac tissues were performed using massively parallel supercomputers, the resources of the National Research Scientific Computing Center, which is supported by the office of Science of the Department of Energy under contract # DE-AC02-OSCH11231.

I express great gratitude for help in editing and formatting of the book to the young generation of the Grayver family under leadership of Dr. E. Grayver and to my PhD student Richard Samade.

Table of Contents

Chapter 1. Introduction

In this chapter, we briefly outline the basic preliminary information about the heart as a pump in the blood circulatory system, its structure, and major component systems. The material presented here is a series of short excerpts from A.C. Guyton and J.E. Hall's "Textbook of Medical Physiology" [1], which is meant to introduce readers without biological background to the basic terminology, definitions and functions of the heart systems as part of the whole organ. We will look at additional physiological information in greater depth in later chapters.

Here we also emphasize that heart arrhythmias and fibrillation continue to be dangerous heart diseases, the mechanisms of which have not been clearly understood until recently. Mathematical modeling and computer simulations are characterized as modern tools for scientific research in this area.

1.1. Heart as a four-chamber pump

1.1.1. Heart function

The heart is a rhythmical, adjustable, muscular pump whose function is to maintain adequate supply of blood at sufficient pressure to meet tissue demands for nutrients and waste removal in all organs of the body (see Fig. 1).

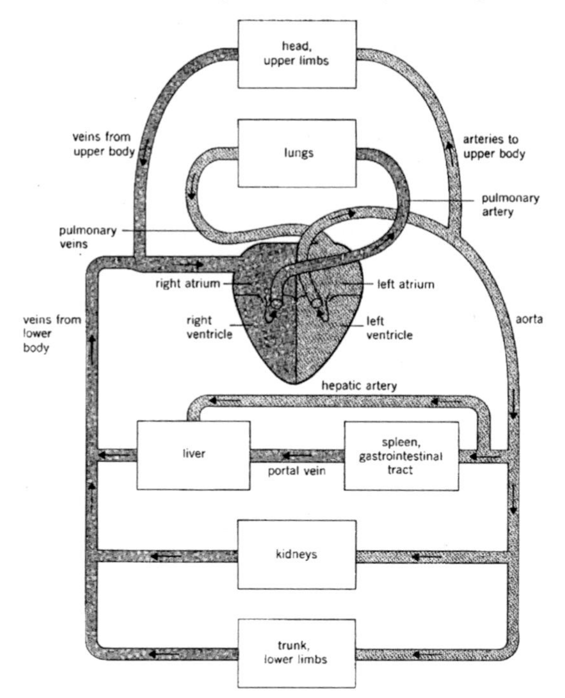

Fig. 1 A schematic representation of the circulatory system [1]

B.Ja. Kogan, *Introduction to Computational Cardiology: Mathematical Modeling and Computer Simulation*, DOI 10.1007/978-0-387-76686-7_1,

1.1.2. Heart structure

The heart is a four-chambered organ supplied with valves to control the direction of blood flow (Fig. 2). It is composed of the basic types of tissue that account for its auto-rhythmicity, conductivity, and contractility.

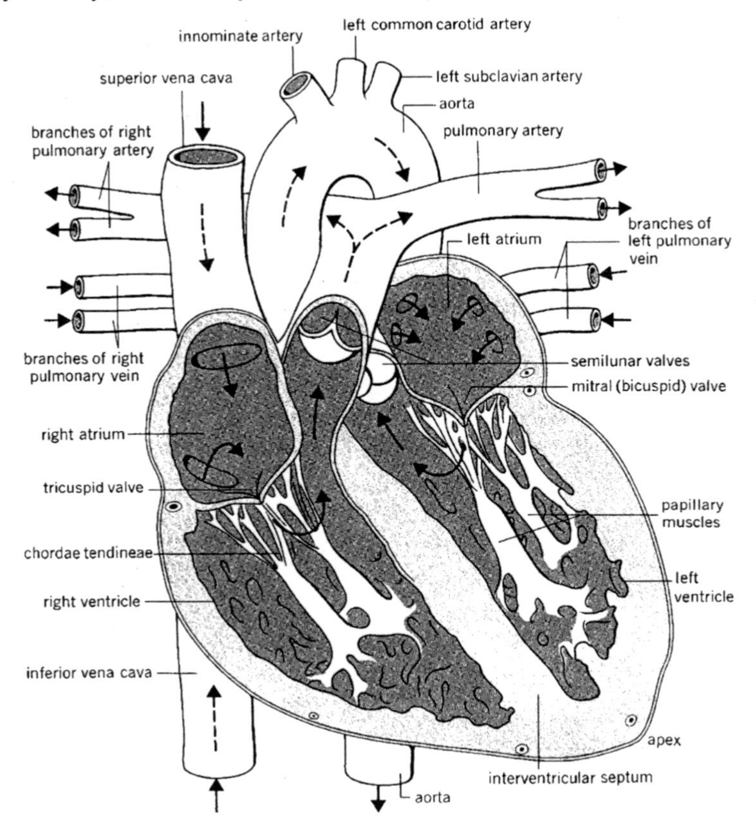

Fig. 2 Diagram of the heart and its major blood vessels. The heart is viewed from the front [1]

1.1.3. Blood supply of the heart-coronary system

The term coronary means literally "a crown," which implies encirclement. The term is apropos since coronary vessels do, in fact, encircle the heart.

The low coronary arteries branch off the aorta just beyond the aortic valve (Fig. 3). Thus, as soon as the blood leaves the left ventricle, it enters the coronary arteries for distribution to the cardiac muscle.

Unfortunately, the two coronary arteries are functionally end arteries. This means that the two vessels are more or less independent of each other. In contrast, in most organs of the body, the various blood vessels meet with one another so that, if one is blocked, blood from the other vessels may still nourish the tissues. But in the heart, if one of the coronary arteries is suddenly occluded, the corresponding part of the heart muscle loses its blood supply.

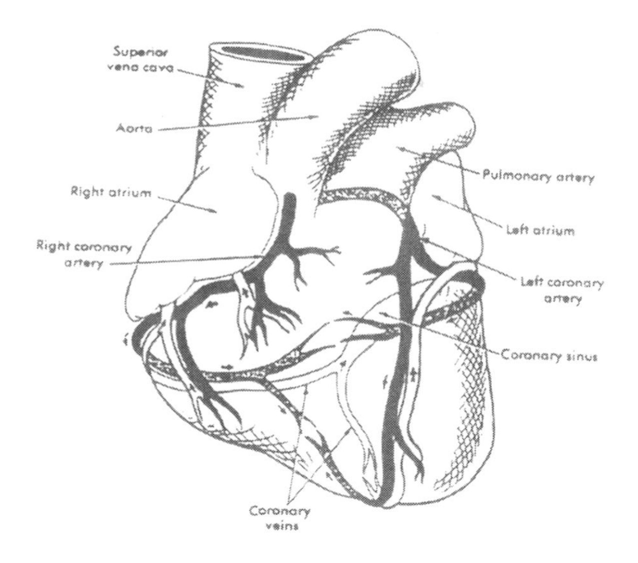

Fig. 3 Anatomy of the coronary circulation [1]

1.1.4. Heart Automaticity

Heart contractions or heartbeats are initiated by a special system called the pacemaker system. This system consists of several nodes that automatically generate excitation pulses at their own different rates (see Table 1). Being connected in the system by atrium tissue, conducting and Purkinje fibers they transfer and distribute along the ventricles' muscles the excitation pulses generated with higher frequency in the system. The location of the each part of pacemaker system in the heart is shown in Fig. 4.

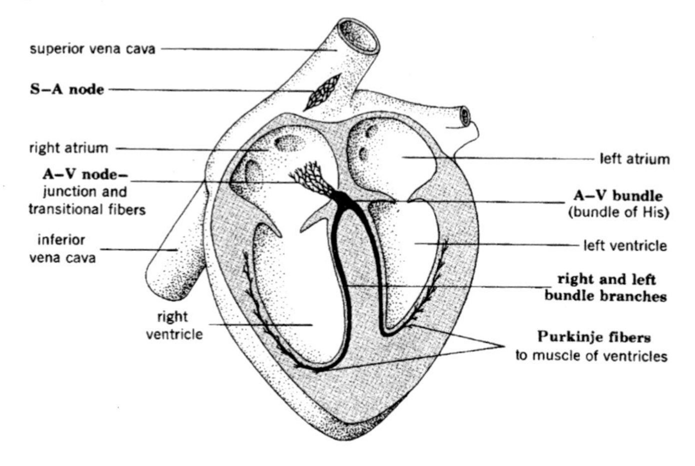

Fig. 4 The locations of the nodal tissue in the human heart [1]. (S-A)- sin-atria node; (A-V)- atria-ventricular node

Table 1. Some characteristics of nodal tissue and cardiac muscle

Area	Velocity of conduction (m/sec)	Inherent rate of discharge (impulses/min)
S – A node	0.05	70-80 (pacemaker)
Atrium muscle	0.3	-
A – V node	0.05-0.1	40-60
A – V bundle	2-4	35-40
Purkinje fibers	2-4	15-40
Ventricular muscle	0.3	-

1.2. Heart systems

Under normal conditions, proper functioning of the heart's major systems (see Fig. 5) supports the normal pump activity. These systems include: heart blood supply, neuro-humoral regulation, metabolism, heart electro-physiology [2, 3], and cardiac muscle contraction system. Distortion in any of these systems may trigger different types of heart diseases including heart arrhythmias leading to ventricular fibrillation.

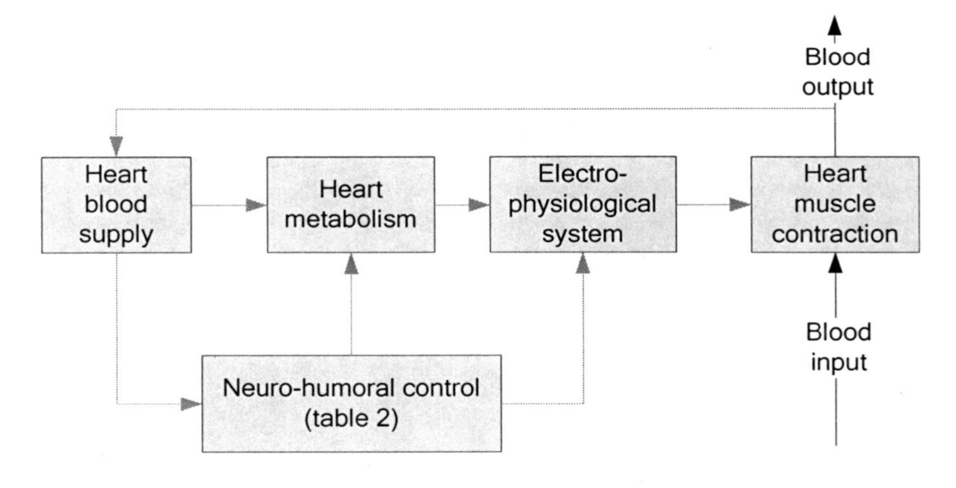

Fig. 5 Heart system (simplified block diagram)

A simplified block diagram of the electrophysiological system is shown in Fig. 6, inside the white square box.

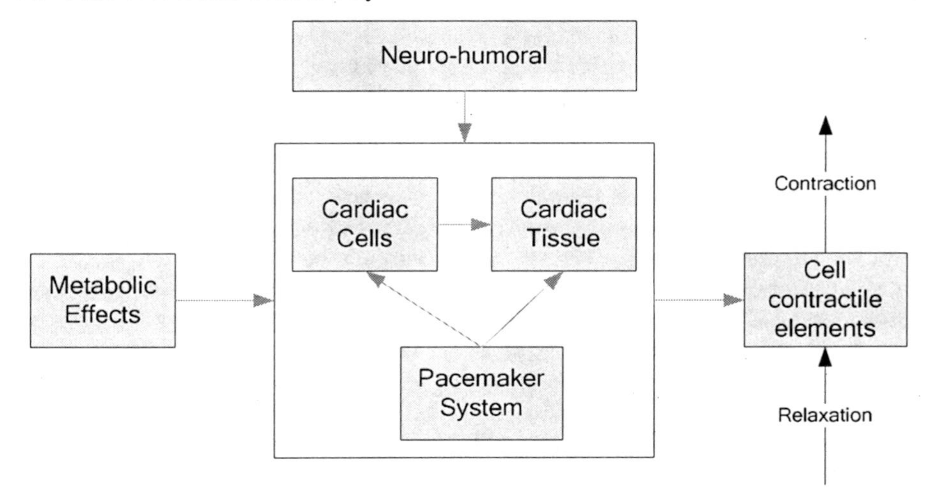

Fig. 6 Electrophysiological system

Nervous control system affects the blood pumping by changing the heart rate and strength of contractions, as described in table 2.

Table 2: Nerve Control of the heart

Type of nerve system	Location of endings	Effect on heart beat rate	Effect on strength of contraction	Mechanism of action: nerve endings release hormones
Sympathetic	Ventricles Both atria (sparsely)	Increase	Increase: Normally 20% Max. 200%	Norepinephrine, which increases membrane permeability to Na ions
Para-sympathetic	SA and AV nodes	Decrease	Decrease by: 10 – 20%	Acetylcholine, which increases membrane permeability to K ions

1.3. Control of Heart Contractility

Heart contractility is the major property of the heart muscle, providing its function as a blood pump. This occurs when special protein cell motors (myofibrils) directly transform chemical energy into mechanical contraction of the heart chamber walls. Contractions have to be periodic (Heart Cycle) and occur in some required sequence in time and space to provide coordinating function of all four chambers of the heart. Special control signals are produced by heart subsystems to perform this function. This is called excitation–contraction coupling.

1.3.1. Excitation-contraction coupling

Excitation-contraction coupling includes three heart subsystem control signals. First the pace-maker system generates the control signals of the basic heart rhythm (70 beats/min). These signals propagate directly through the atria and indirectly through the ventricular (via Giss fibers and Purkinje cardiac cells) and excite the atria and ventricular cell membranes (appearance of membrane depolarization and subsequent AP) in the required space distribution and timing. The AP may be considered as a second control signal on the way to cell contraction. It causes the opening of the Ca cell membrane channels, and in response the Ca releases from intracellular sarcoplasmic reticulum stores, providing the intracellular Ca concentration changes required for myofibril contraction and then relaxation (see [6] for details). Thus, Ca dynamics may be considered as a third control signal on the way to excitation-contraction coupling. A block diagram of this three-step control system is illustrated schematically in fig. 7.

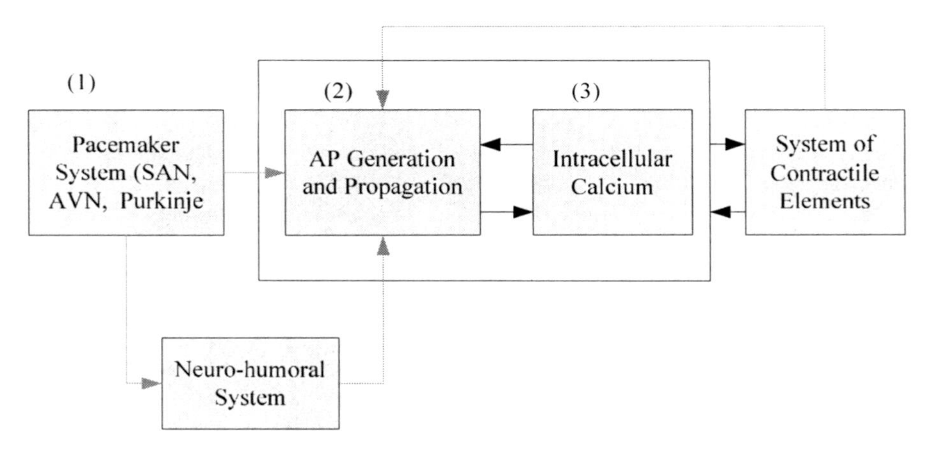

Fig. 7 Simplified block diagram of the heart control

1.4. Heart Fibrillation and Sudden Cardiac Death

Sudden cardiac death takes the lives of about a half million people every year in the US alone. As it is widely recognized now, the main cause of this death is ventricular fibrillation (VF), the severe distortion of cardiac muscle contraction rhythm (see fig. 8 and fig. 9), leading to full loss of the heart's pumping function. VF may arise in both originally healthy hearts, and ischemic hearts. Both cases are considered in modeling and computer simulation.

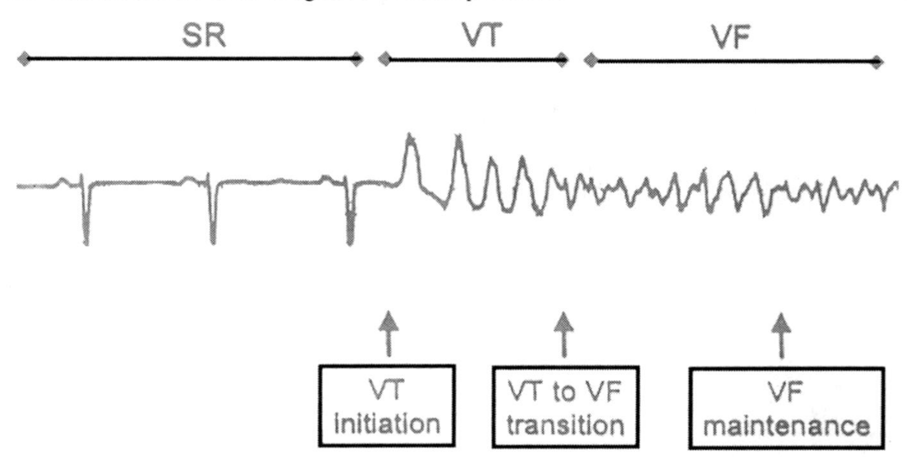

Fig. 8 Stages of sudden cardiac death [4] (EKG record) (SR – sinus rhythm, VT – ventricular tachycardia, VF- ventricular fibrillation)

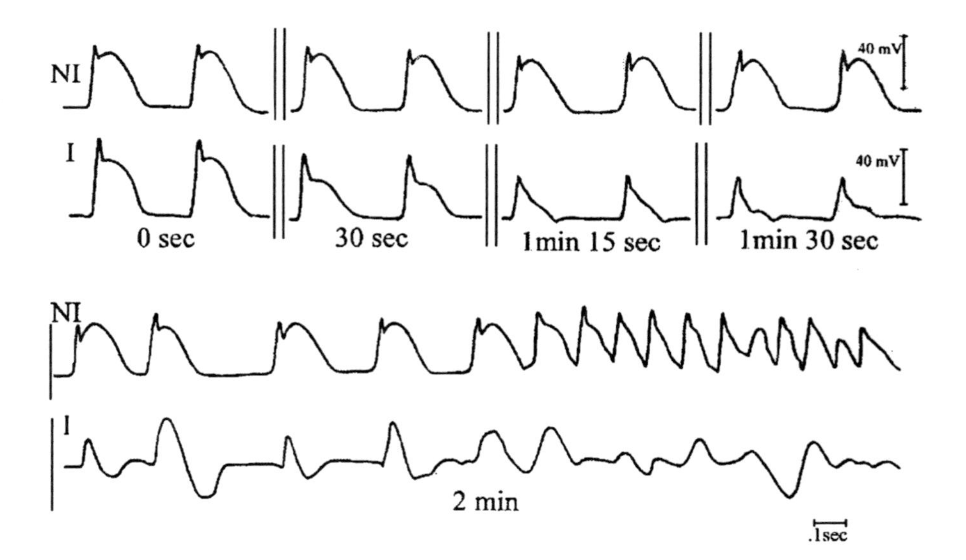

Fig. 9 The time course of monophasic AP dog's heart in ischemic (I) and non-ischemic (NI) zones after coronary artery occlusion (CAO) [5].

1.5. The methods for investigation of heart processes

Two complementary main approaches are now widely used to study the behavior of the heart's systems under normal and pathological conditions. One is based on experimental investigations performed on different biological levels (cellular, molecular, tissue, separate organ, and whole organism); the other uses mathematical modeling and computer simulations. Particular results of both approaches are shown in Fig. 10.

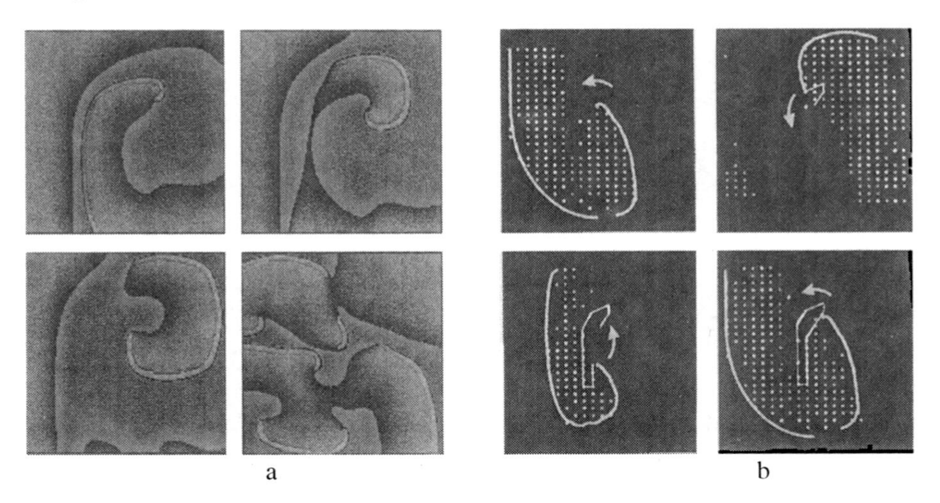

Fig.10 Examples of spiral wave: a. observed in computer simulation [7], and b. spiral wave in experimental record [8].

It is appropriate to emphasize here the fundamental role and importance of mathematical methods in supporting these approaches and generalizing the results (see [9]). Both approaches have their drawbacks, advantages and limitations, which we will discuss later. Only cooperative use of these approaches in combination with broad use of mathematical methods can provide reliable and useful results.

This book is focused on the modeling and computer simulation of electrophysiological processes in myocardium. The development of a mathematical model of the whole heart (as a hierarchy of the particular model of heart systems and subsystems) is the subject of a special project, which is called the *physiome of the heart* [10].

1.6. Role of mathematical modeling and computer simulation in investigating the heart processes

Mathematical modeling and computer simulations are powerful methods used for research in almost all areas of science and technology. They possess the predictive ability and, in conjunction with the experimental method, facilitate the discovery of new phenomena and understanding of its mechanisms. These methods were also widely used for the last fifty years for solving an array of both general and specific problems in cardiology. Some of these include:

1. Investigation of electrophysiological mechanisms under normal and pathological conditions (without and with intracellular Ca dynamics)

2. Validation of hypotheses proposed during investigations of ventricular fibrillation appearance (ectopic activity, re-entry of excitation, role of arrhythmias and subsequent changes in repolarization cell processes, etc.)
3. Creation of a theory of stationary spiral wave propagation
4. Investigation of mechanisms of heart defibrillation and reasons for defibrillation failure
5. Investigation of the effect of protein mutations in cardiac cell channels on action potential generation and propagation
6. Study of the effect of anti-arrhythmic and anti-fibrillatory drugs
7. Creation of a whole heart mathematical model with three-dimensional geometry.

Not all of these problems are fully solved, and some, such as genetic effects, the role of mechanical contraction in the myocardial electrical activity, and others await a solution.

The additional advantages of mathematical modeling and computer simulation are the possibility of gaining insight into some internal processes and/or parameter changes that cannot be directly observed in the course of physiological experiments. They also allow us to isolate each process under investigation and study the effects of different factors independently (parameters, stimulation, initial conditions, etc).

Validation of the mathematical model and computer simulation results is of primary importance.

1.7. References

1. Guyton, A.C. and J.E. Hall, *Textbook of Medical Physiology*. 9th ed. 1996, Philadelphia: W.B. Saunders.
2. Hoffman, B.F. and P.F. Cranefield, *Electrophysiology of the Heart*. 1960, New York: McGraw-Hill.
3. Zipes, D. and J. Jalife, eds. *Cardiac Electrophysiology: From Cell to Bedside*. 2nd ed. 1995, W.B. Saunders: Philadelphia.
4. Weiss, J.N., *Transition of Sinus Rhythm to VT to VF*, Unpublished.
5. Rajskina, M.E., *Ventricular Fibrillation and Sudden Coronary Death*. 1999: Kluwer Academic Publishers.
6. Bers, D.M., *Excitation-Contraction Coupling and Cardiac Contractile Force*. 2nd ed. 2001, Norwell, MA: Kluwer Academic Publishers.
7. Garfinkel, A., P.-S. Chen, D.O. Walter, H.S. Karagueuzian, B. Kogan, S.J. Evans, M. Karpoukhin, C. Hwang, T. Uchida, M. Gotoh, O. Nwasokwa, P. Sager, and J.N. Weiss, *Quasiperiodicity and chaos in cardiac fibrillation*. J Clin Invest, 1997. **99**: 305-314.
8. Davidenko, J.M., A.V. Pertsov, J.R. Salomonsz, W. Baxter, and J. Jalife, *Stationary and drifting spiral waves of excitation in isolated cardiac muscle*. Nature, 1992. **355**: 349-351.
9. Keener, J. and J. Sneyd, *Mathematical Physiology*. 2nd ed. 2001: Springer-Verlag.
10. Bassingthwaighte, J.B., *Toward modeling the human physiome*. Adv Exp Med Biol, 1999. **382**: 331-339.

Chapter 2. Mathematical Modeling and Computer Simulation

Once upon a time, man started to use models in his practical activity. Modeling continues to play a very important role in studying natural phenomena and processes as well as helping to create modern engineering systems. Additionally, modeling is used in biology and medicine to find the mechanisms of function and malfunction concerning the organs of living organisms at both the micro and macro level.

Generally, a model has been defined [1] as the reconstruction of something found or created in the real world, a simplified representation of a more complex form, process, or idea, which may enhance **understanding** and facilitate **prediction**. The object of the model is called the **original,** or **prototype system**.

The model and the original may have the same physical nature; such models are called **physical models**. Correct physical models must satisfy the criteria of similarity, which include not only the conditions of geometrical similarity but also similarity of other characteristics (for example: temperature, strength of electromagnetic field, etc.). Physical models have been widely used in engineering and biomedicine. Examples include the testing of various civil constructions for seismic stability, testing the aero-dynamic characteristics of new aircraft and rockets in wind tunnels, and experimental studies on animals (organ, tissue, and cell) considered as a prototypes for human beings.

However, in scientific research this type of modeling studies is complemented with another modeling approach, which is based on the development of mathematical descriptions of the behavior of the prototype system under investigation. These descriptions are called mathematical models. The results are expected to be obtained by using existing mathematical methods (which give the solution in closed form mostly for very simplified cases) or by computer simulation using powerful serial or parallel supercomputers.

In this chapter we present definitions and terminology, classification of mathematical models, general assumptions accepted in mathematical modeling, and some considerations about mathematical models of direct analogy (see also Appendix) and computer simulations.

2.1. Mathematical modeling

The place of mathematical modeling among the other methods of scientific investigation [2] is shown schematically in Fig. 1.

B.Ja. Kogan, *Introduction to Computational Cardiology: Mathematical Modeling and Computer Simulation*, DOI 10.1007/978-0-387-76686-7_2,

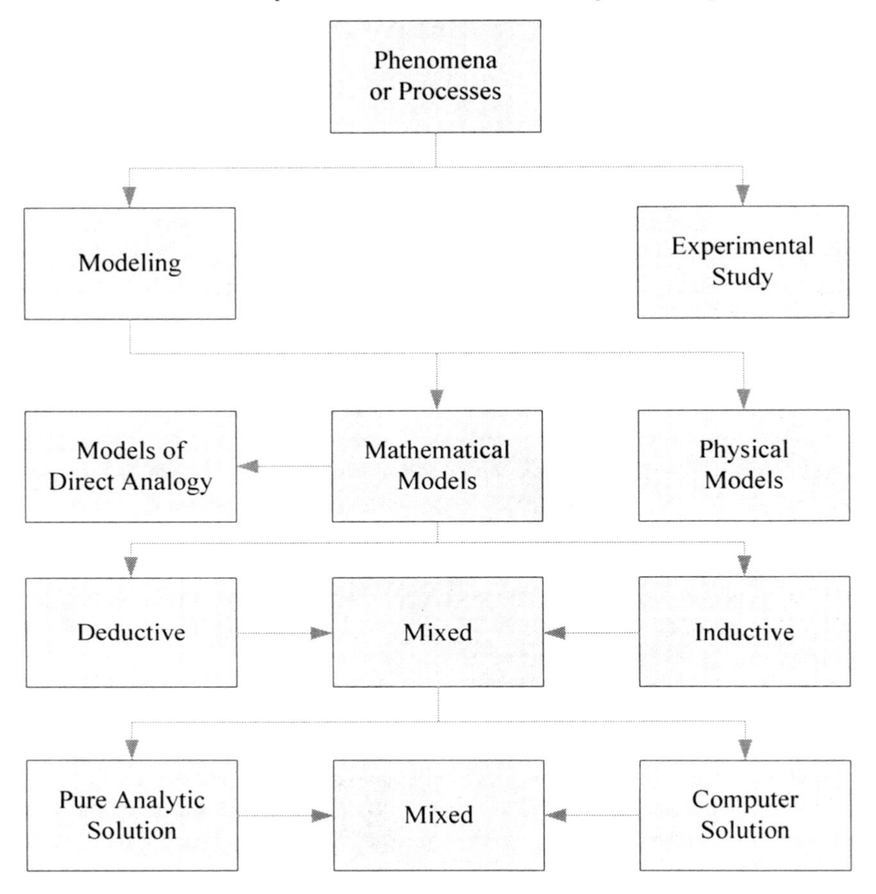

Fig. 1 Schematic representation of different modeling approaches

Mathematical models represent a mathematical description of the original, based on known general laws of nature (First Principles) and experimental data. The well-known fact that the systems of different physical natures have the same mathematical descriptions led to a special type of mathematical models: models of direct analogy. The tremendous advancements in computer hardware and software stimulated the wide use of mathematical models, especially because most of the new problems, particularly in physiology, are nonlinear and, thus, their solutions cannot be obtained analytically in closed form.

Mathematical modeling facilitates the solution of three major problems for a prototype system: analysis, synthesis and control. The characteristic of these problems (see [3]) is given in Fig. 2 and Table 1.

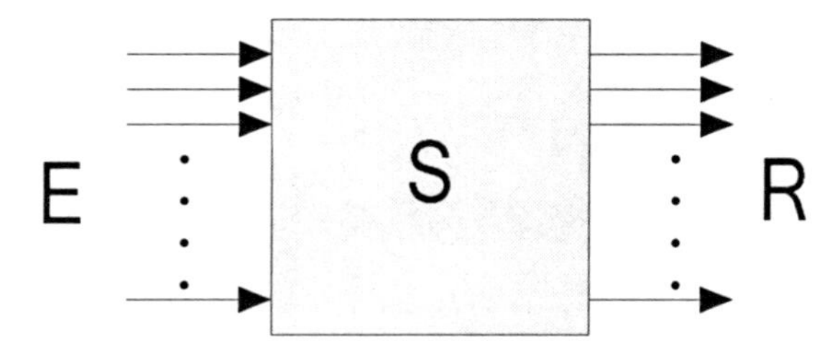

Fig. 2 The cause-and-effect relation between excitation, E, and, response, R as they relate to the system S

Problems can be classified according to which two of the items E, S, R are given and which is to be found. E represents excitations, S the system, and R the system's responses.

Table 1. General classification of the problems

Type of Problem	Given	To find
Analysis (direct)	E, S	R
Synthesis (design identification)	E, R	S
Instrumentation (control)	S, R	E

The analysis problem is sometimes referred to as the direct problem, whereas the synthesis and control problems are termed as inverse problems. A direct problem generally has a unique solution. For example, if the Noble mathematical model of Purkinje fiber [4] is used, we obtain only one action potential shape in response to a specified stimulus for given cell parameters. In contrast, the inverse problem always gives an infinite number of solutions. To find a single solution additional conditions and constraints must be specified separately. An example of this is found in the modeling of Ca^{2+} induced Ca^{2+} release mechanisms from the cardiac cell sarcoplasmic reticulum (SR).

The spectrum of mathematical models can be constructed based on our prior knowledge of the prototype system (see Fig. 3 taken from [3] and reflecting the situation in the year 1980). The darker the color, the more restricted our knowledge about the system, and the more qualitative the simulation results. As our knowledge of prototype systems progresses, some parts of this spectrum became brighter and the possibility of obtaining quantitative results increases.

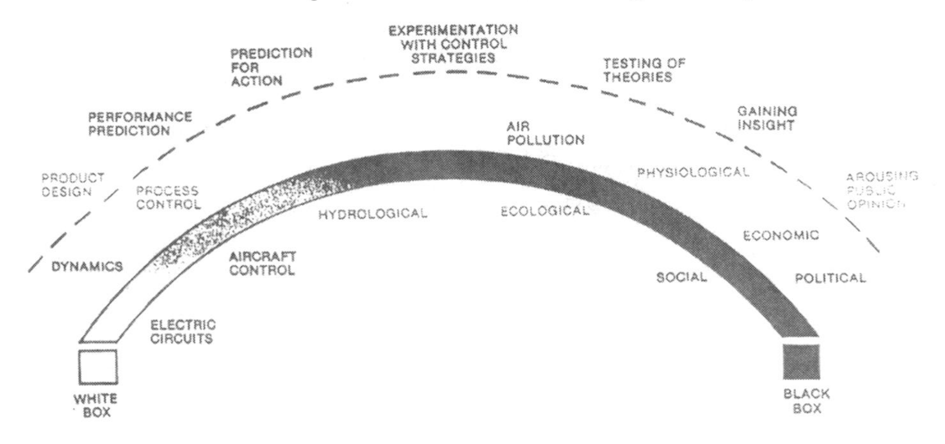

Fig. 3 Motivations for modeling [3] showing the shift from quantitative models (light end of the spectrum) to qualitative models (dark end)

2.1.1. Deductive, inductive and combined mathematical models

In cases when there is enough knowledge and insight about the system, the deductive approach is used for model formulation. **Deduction** derives knowledge from known principles in order to apply to them to unknown ones; it is reasoning from the general to the specific. The deductive models are derived analytically (from first principles), and experimental data is used to fill in certain gaps and for validation. The alternative to deduction is induction. Generally, **induction** starts with specific information in order to infer something more general. An induction approach in biomedicine is fully based on experimental observations and has led to the development of numerous phenomenological models (e.g. Wiener and Rosenbluth [5], Krinski [6], Moe [7] models of the cardiac cell). In most practical modeling situations of the heart processes, both deductive and inductive approaches are required. The gate variable equations introduced by Hodjkin-Huxley [8], derived from the cell-clamp experiments, are an example of an inductive approach, whereas the application of Kirchoff's law to the current balance through the cell membrane is an indicator of the deductive approach used in formulating the action potential models for nerve and heart cells.

Using induction, we must accept the possibility that the model might not be unique and its predictions will be less reliable than when the model is purely deductive. Consequently, such a model will have less **predictive validity**; defined as the ability of the model to predict the behavior of the original system under conditions (inputs) which are different from that used when the model was originally formulated. Most of the mathematical models in biology are semi-phenomenological. This means that part of the model derives from first principles (the laws of conservation of matter and energy) and the rest represent the appropriate mathematical interpretation of experimental findings.

2.1.2. General assumptions used in mathematical modeling

Some simplified assumptions of general character are used in formulation of mathematical models. These assumptions relate to the general properties of the original system or phenomena under investigation:

1. ***Separability*** makes it possible to divide the entire system into subsystems and study them independently (with the possibility of ignoring some interactions). For example, typically AP models do not include cardiac cell metabolic processes. Practically, they remain unchanged during the time course of many cardiac cycles (changing over different time scales).
2. ***Selectivity*** makes it possible to select some restricted number of stimuli, which affect the system. The excitable membrane, for example, can be excited by current stimulus, changing the concentration of chemical substances inside the cell and changing the cell temperature.
3. ***Causality*** makes it possible to find cause and reason relationships. It is not enough to observe that variable 'y' always appears after variable 'x'. There is a possibility that they both are the result of the common reason-variable 'u'.

2.1.3. Mathematical Models of direct analogy

Let us consider, as examples, the mathematical models of two prototype systems with different physical natures. The first is an electrical lumped R, L, C circuit and the second is a mechanical mass, spring system with damping. Both are shown in Fig. 4.

The electrical circuit serves here as a mathematical model of direct analogy for mechanical systems and vice versa. With the development of powerful computers the role of direct analogy models becomes negligibly small. Nevertheless, historically, the FitzHugh-Nagumo simplified AP model, which is still widely used today [4], was derived for nerve cell study as a direct analogy for the Van der Pol equation of relaxation oscillation (see Chapter 5).

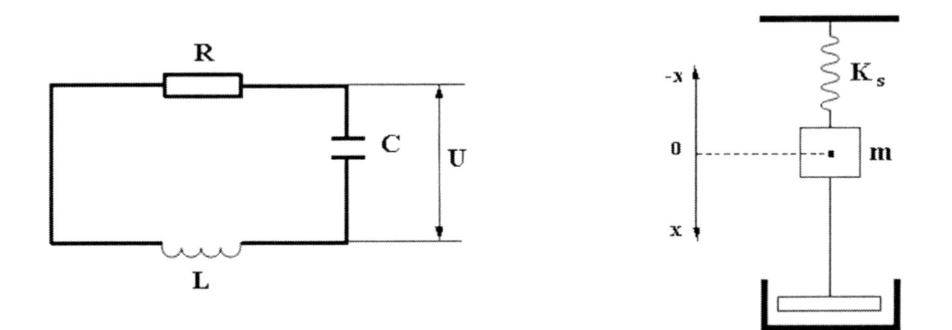

Fig. 4 Schematic diagram of electrical and mechanical oscillators.

We will assume that the capacitor was initially charged to the voltage $U_c(0) = U_{co}$ and mass m was initially displaced (from the equilibrium position $x = 0$) by the value $x(0) = x_0$ and released with $(dx/dt)_{t=0} = 0$. We also suppose that these perturbations are small enough to consider that the parameters of the systems remain constant.

Kirhoff's law for electrical circuits and Newton's law for mechanical systems give respectively:

a. Balance of voltages in an electrical circuit:

$$U_L + U_R + U_C = 0; \text{ where: } U_L = L\frac{di}{dt}; \qquad U_R = iR; \qquad i = C\frac{dU_C}{dt}$$

Thus,

$$\frac{d^2U_C}{dt^2} + \frac{R}{L}\frac{dU_C}{dt} + \frac{1}{LC}U_C = 0 \tag{1}$$

b. Balance of forces in a mechanical system:

$$F_a + F_d + F_s = 0, \text{ where: } F_a = m\frac{d^2x}{dt^2}; \qquad F_d = k_d\frac{dx}{dt}; \qquad F_s = k_s x;$$

So,

$$\frac{d^2x}{dt^2} + \frac{k_d}{m}\frac{dx}{dt} + \frac{k_s}{m}x = 0 \tag{2}$$

The coefficients: $R/L = 2\alpha_e$ in (1) and $k_d/m = 2\alpha_m$ in (2) are the damping ratios; coefficients $1/LC = (\omega_0^2)_e$ and $k_s / m = (\omega_0^2)_m$ represent the squares of natural angular frequencies for systems (1) and (2) respectively.

The solutions of (1) and (2) depend on the ratio $\alpha_e / (\omega_0)_e$ and $\alpha_m /(\omega_0)_m$ correspondingly.

For initial conditions:

$U_c(0) = U_{C0}$, $(dU_c/dt)_{t=0} = 0$ and $x(0) = x_0$ $(dx/dt)_{t=0}=0$ and when parameters are such that $\alpha_e < (\omega_0)_e$ and $\alpha_m < (\omega_0)_m$ we get:

$$U_C = U_{C_0} e^{-\alpha_e t} \cos\omega_e t; \qquad x = x_0 e^{-\alpha_m t} \cos\omega_m t \tag{3}$$

Here, $\omega_e = \sqrt{(\omega_0)_e^2 - \alpha_e^2)}$ and $\omega_m = \sqrt{(\omega_0)_m^2 - \alpha_m^2)}$

If $\alpha_e = \alpha_m$ and $\omega_{0e} = \omega_{0m}$, then $U_c(t) / U_{c0} = x(t) / x_0(t)$ and we can study the behavior of a mechanical system using an electrical circuit where it is easier to perform the measurements and change the system parameters.

Using this example it is possible to notice that both systems, when using the appropriate initial conditions, are mathematically described by the same differential equation:

$$\frac{d^2u}{dt^2} + 2\alpha\frac{du}{dt} + \omega_0{}^2 u = 0 \tag{4}$$

with appropriate initial conditions.

This equation represents the mathematical model for second order linear dynamic systems, independently of the physical nature of state variable u. In Table 2 we demonstrate the predictive ability of this model.

Table 2 The predicted behavior of a linear oscillator based on the mathematical model of direct analogy

	$\alpha > 0$	$\alpha < 0$	$\alpha = 0$
$\alpha^2 < \omega^2$	Sinusoidal oscillations with decreasing amplitude	Sinusoidal oscillations with increasing amplitude	sinusoidal oscillations with constant amplitude
$\alpha^2 > \omega^2$	Aperiodic process with decreasing amplitude	Aperiodic with increasing amplitude	

2.1.4. Relaxation oscillations

Van Der Pol [10] discovered relaxation oscillations when he investigated the problem of stabilization of the amplitude of a carrier signal generated to broadcast radio translations. For this purpose he proposed the introduction of nonlinear positive damping proportional to the square of oscillation amplitude in addition to negative damping (α<0) in the second order oscillator equation (4). The equation (4) with this modification attains the form:

$$\frac{d^2u}{dt^2} + 2\alpha(1-\beta u^2)\frac{du}{dt} + \omega_0{}^2 u = 0 \tag{5}$$

Here β is a coefficient usually chosen equal to one.

This is the Van Der Pol equation. Its solution for $\alpha^2 < \omega_0{}^2$ and $\alpha < 0$ is shown in Fig.5. Each time the amplitude u becomes more or less than unity the sign of the damping ratio changes respectively stabilizing the amplitude of oscillation.

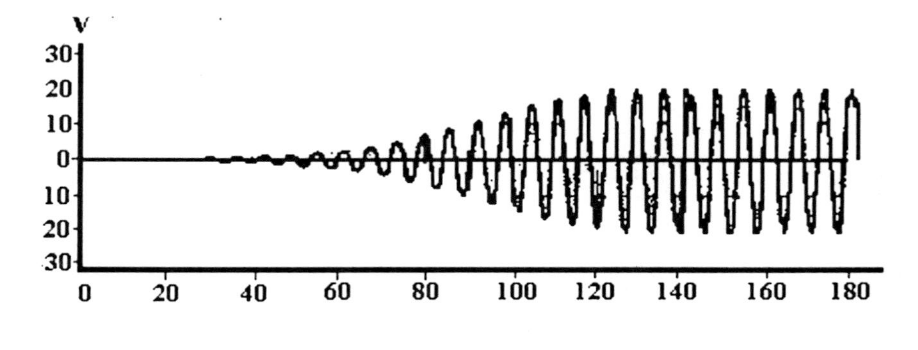

Fig. 5 The solution of Van Der Pol equation for $\varepsilon = \dfrac{\alpha}{\omega_0} = 0.1$

The relaxation oscillations were discovered as a solution of equation (5) for $\alpha < 0$ and when $\alpha^2 >> \omega_0^2$ (see Fig. 6)

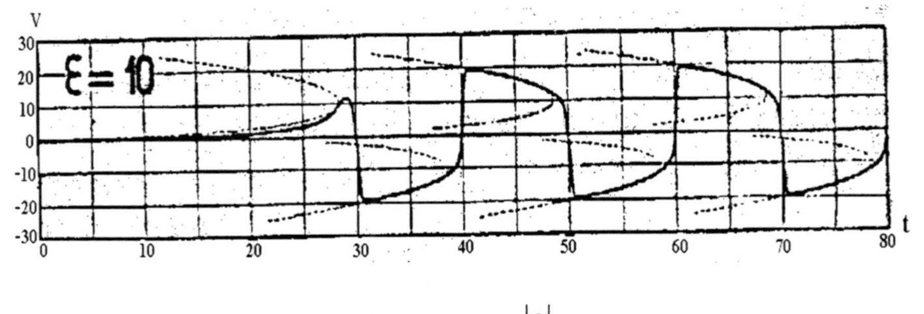

Fig. 6 Relaxation oscillations, $\varepsilon = \dfrac{\alpha}{\omega_0} = 10$. $T_{rel} = \dfrac{|\alpha|}{\omega^2} = RC$ approximately gives the period of the relaxation oscillations.

Van Der Pol proposed using the sequentially connected relaxation oscillators as a model of the heart pacemaker system [10]. For this purpose each relaxation oscillation generator in the system is adjusted to the frequency of the corresponding pacemaker system node. The discovery of relaxation oscillations and the development of the phase-plan approach in the analysis of nonlinear dynamic systems facilitated the development of simplified nerve and heart cells models (see Chapter 5 for details)

2.1.5. Validation of mathematical models

Mathematical model validation involves the comparison of computer simulation results with those obtained on a real prototype of the simulated object, assuming the digital computer implementation introduces negligible additional errors. Model identification theory and methods have been developed for most linear and quasi-linear dynamical systems (in engineering and some in biology). These methods can be used to identify the parameters [11] and even structure of the model without and with the presence of noise [12]. One of the possible block diagrams of mathematical model validation and identification is presented in Fig. 7.

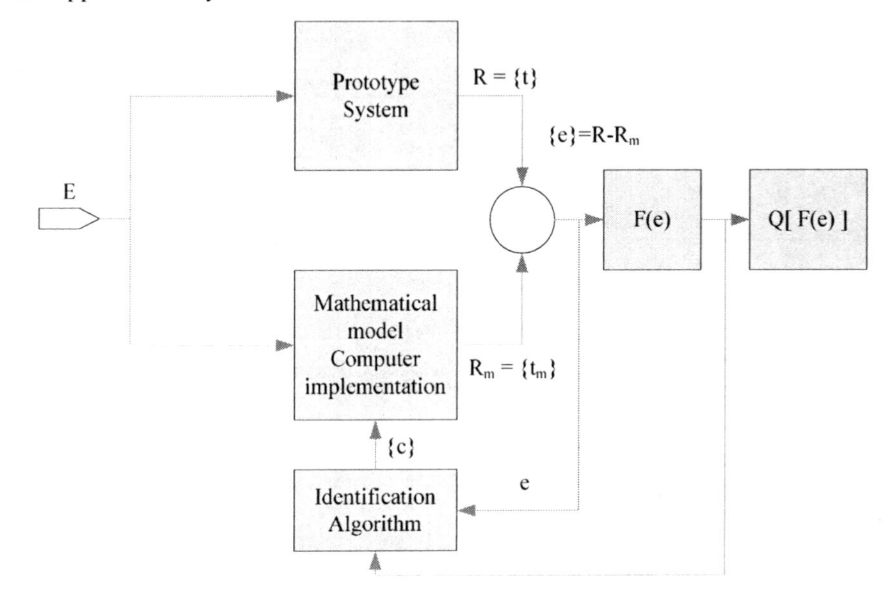

Fig. 7 Simplified block diagram for validation and identification of a mathematical model. E is a vector of chosen input excitation, R is the measured response, c is a vector of control parameters of the model, and {e} is a vector of mathematical model error. F{e} is a chosen error function (typically, $F\{e\}=e^2$). Q[F{e}] is a criterion of identification quality (typically mean square)

Unfortunately, most biological systems are significantly nonlinear dynamical systems (nerve, heart, vascular and skeletal muscle systems etc), which cannot be reduced to linear or quasi-linear models without loss of their major functions. Moreover, for these systems even the most advanced experimental technology cannot provide the necessary data not only for full verification but also for formulation of some phenomenological part of the mathematical model (for example, there is no experimental data to formulate the mathematical model of spontaneous Ca release from SR). In these cases, some plausible hypothesis is usually formulated and the model predictions are considered correct until new contradictory experimental data is obtained. Similar situations have been encountered throughout the history of studying different natural phenomena.

2.2. Appendix: Lilly-Bonhoeffer Iron Wire Model

William Ostwald (1900) [13] was the first to notice that iron wire in nitric acid exhibits an electrochemical surface phenomenon quite similar to the action potential in nerves. Later the iron wire model was investigated experimentally by Lillie [14] and theoretically by Bonhoeffer [15]. The one-dimensional iron wire model is a mathematical model of direct analogy for nerve pulse propagation and is shown in Fig. 8.

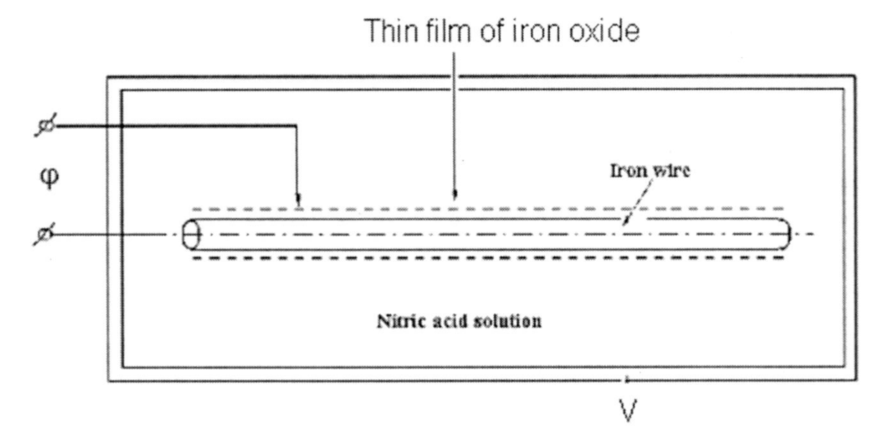

Fig. 8. One-dimensional iron wire model. $\varphi - electricalpotential$

The iron wire, IW, is immersed in the vessel, V, filled with nitric acid of some concentration (electrolyte). The iron wire is covered with a thin film of iron oxide, shown by the dotted line in Fig. 8. After an application of a suprathreshold current stimuli, the difference of potential between the iron wire and electrolyte, φ, rises so that the thin film of iron oxide is destroyed at that place. Then, this potential accompanied by the destruction of the thin film begins to propagate toward the two ends of the wire, resembling the propagation of a nerve pulse along the nerve fiber.

The mathematical model can be derived from the current balance in the system:

$$C\frac{\partial\varphi}{\partial t} = i_f + i_a + i_{na} + i_{Iw} + i_{st} + \frac{1}{R}\frac{\partial^2\varphi}{\partial x^2}$$

Here:

φ the difference of potential between the iron wire and the electrolyte

C capacitance of a double layer

T time

i_f thin film current $i_f = \begin{cases} (1-\alpha)k_2(\varphi) & \varphi. \geq \varphi_{th} \\ \alpha k_2(\varphi) & \varphi < \varphi_{th} \end{cases}$

α degree of activation $\frac{\partial\alpha}{\partial t} = -\frac{1}{Q}i_f$

Q electrical charge per unit of film surface

i_{IW} iron wire current $i_{IW} = k_1(\varphi)\alpha$

i_{na} nitric acid current; $i_{na} = [NA]k_3(\varphi)$

[NA] the concentration of nitric acid near the surface of wire

$$\frac{\partial [NA]}{\partial t} = D \frac{\partial^2 [NA]}{\partial y^2}$$

I_{st} stimuli current
R the longitudinal specific resistance of electrolyte
D diffusion coefficient

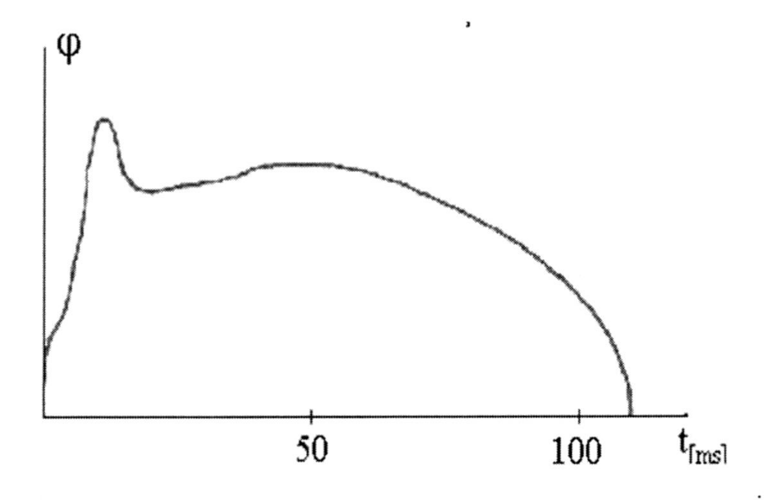

Fig. 9 Shape of propagated potential

A grid of iron wires (see Fig. 10*a*) supports propagation of 2D waves. The first publication of this experimental system [16] exhibited both circular waves radiating from a point source of excitation and spiral waves rotating loosely about one endpoint of the wavefront (Fig. 10*b*). Figure 10a shows a 26×26 grid of iron wires (30cm × 30cm). Figure 10b shows pencil tracing at 1/8s intervals (left to right, then down arrow) taken from a photo of an iron wire grid when stimuli were introduced at S_1, S_2, S_3, and S_4. Spontaneous activity persisted for a while in the form of waves irregularly pivoting about moving points [16].

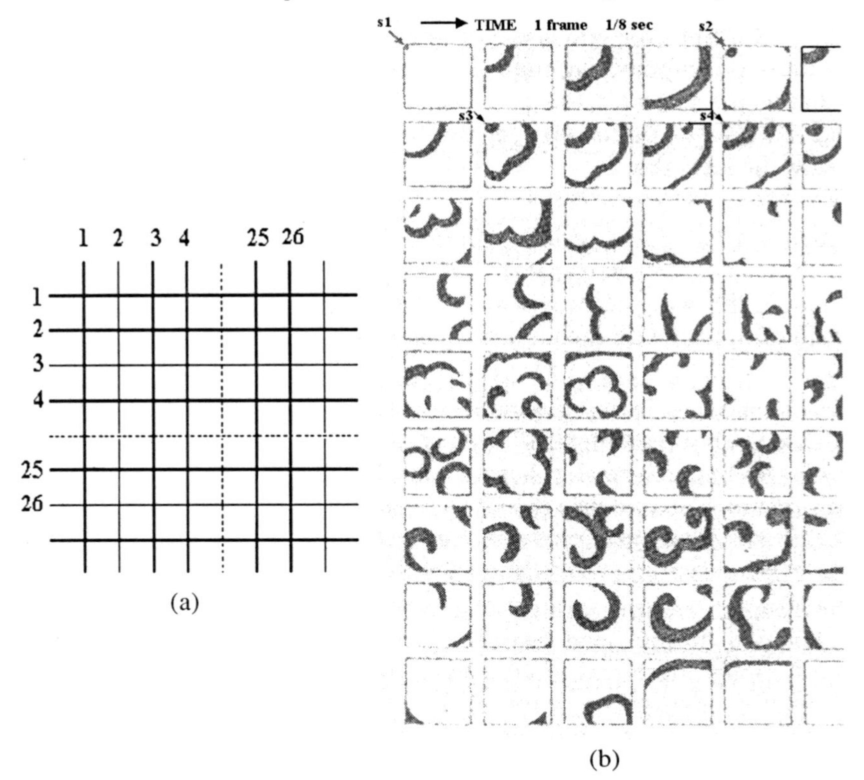

Fig. 10. The 2D iron wire model. *a*: grid of iron wires. *b*: circular wave propagation.

A two dimensional closed surface (a ten-inch iron sphere) behaves in many ways like a human heart, even "fibrillating" when made too excitable or stimulated too frequently (see Smith and Guyton, [17]). This type of model was vigorously investigated for decades (see [18, 19] for a review). Fortunately for many in the West, this remarkable and thorough study published only in Japanese became known to English speakers thanks to a book published by the late outstanding scientist A. Winfree [20].

2.3. References

1. Bekey, G.A., Models and reality: some reflections on the art and science of simulation. Simulation, 1977. **29**: 161-164.
2. Kogan, B.Y. and I.M. Tetelbaum, Modeling techniques in scientific research. Automation and Remote Control, 1979. **40**: 917-924.
3. Karplus, W.J., The spectrum of mathematical models. Perspectives in Computing, 1983. **3**: 4-13.
4. Noble, D., A modification of the Hodgkin-Huxley equations applicable to Purkinje fiber action and pace-maker potentials. J Physiol, 1962. **160**: 317-352.
5. Wiener, N. and A. Rosenblueth, The mathematical formulation of the problem of conduction of impulses in a network of connected excitable elements, specifically in cardiac muscle. Arch. Inst. Cardiol. Mexico, 1946. **16**: 205-265.
6. Krinsky, V.I., Fibrillation in excitable media. Problems in Cybernetics, 1968. **20**: 59-80.

7. Moe, G.K., W.C. Rheinboldt, and J.A. Abildskov, A computer model of atrial fibrillation. Am Heart J, 1964. **67**: 200-220.
8. Hodgkin, A.L. and A.F. Huxley, A quantitative description of membrane current and its application to conduction and excitation in nerve. J Physiol, 1952. **117**: 500-544.
9. Moskalenko, A.V., A.V. Rusakov, and Y.E. Elkin, A new technique of ECG analysis and its application to evaluation of disorders during ventricular tachycardia. Chaos, Solitons & Fractals, 2008. **36**: 66-72.
10. van der Pol, B. and J. van der Mark, The heartbeat considered as relaxation oscillations, and an electrical model of the heart. Archives Neerlandaises Physiologe De L'Homme et des Animaux, 1929. **XIV**: 418-443.
11. Rao, V.S.H. and N. Yadaiah, Parameter identification of dynamical systems. Chaos, Solitons & Fractals, 2005. **23**: 1137-1151.
12. Tzypkin, Y.Z., Information Theory of Identification. 1995, Moscow: Nauka-Fizmatlit.
13. Ostwald, W., Periodische erscheinungen bei der auflosung des chrom in sauren. Zeit Phys Chem, 1900. **35**: 33-76 and 204-256.
14. Lillie, R.S., The electrical activation of passive iron wires in nitric acid. J Gen Physiol, 1935. **19**: 109-126.
15. Bonhoeffer, K.F., Modelle der nervenerregung. Naturwissenschaften, 1953. **40**: 301-311.
16. Nagumo, J., R. Suzuki, and S. Sato. Electrochemical active network. in Notes of Professional Group on Nonlinear Theory of IECE. 1963. Japan.
17. Smith, E.E. and A.C. Guyton, An iron heart model for study of cardiac impulse transmission. Physiologist, 1961. **4**: 112.
18. Suzuki, R., Electrochemical neuron model. Adv Biophys, 1976. **9**: 115-156.
19. MacGregor, R.J. and E.R. Lewis, Neural Modeling: Electrical Signal Processing in the Nervous System. 1977, New York: Plenum.
20. Winfree, A.T., The Geometry of Biological Time. 1980, New York: Springer-Verlag.

Chapter 3. Electrophysiological and Electrochemical Background

In this chapter we present some selected elementary knowledge concerning both the electro-physiology and electro-chemistry of the heart (for details see [1, 2, 3, 4]). This information covers the structures and properties of cardiac cells and tissues, required for understanding the formulation of the mathematical models of action potential (AP) generation and propagation. The basic terminology and definitions are introduced as is required by the discussed topics. Additional physiological information and definitions will be introduced in the following chapters as soon as they become necessary.

The basic distinctive features of cardiac cells and tissues are the properties of excitability and contractility. Excitability is defined as the ability of cardiac cells and tissues to respond to an over-threshold stimulation by generating and propagating electrical pulses – action potentials (APs). Contractility is the ability of cell contractile elements to shorten the cell length in response to an increase in intracellular Ca concentration in the process of developing AP (depolarization phase of membrane potential)

Heart excitable tissues are composed of several excitable cells (with their individual energy sources), which have some common features but differ in other respects by structure and function depending on their location in the heart. Thus Atrium and Ventricle cells exhibit both excitability and contractility properties while the cells of the pace-maker system, similar in their functions to nerve fibers (axons), exhibit only excitability.

In living beings similar properties are found in the cells of both skeletal and smooth muscles. We encounter the properties of excitability also in chemistry (oscillation reactions), electronics, crystallography, plasma systems, etc.

3.1. Cardiac cell and cardiac muscle structures

Heart muscle is formed from bundles of heart fibers consisting of cardiac cells interconnected along their axial and transversal directions as it is shown in Fig. 1.

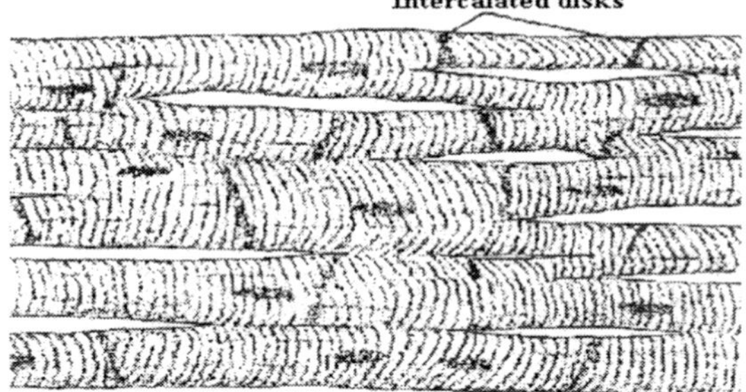

Fig 1 The structure of cardiac muscle [5]

B.Ja. Kogan, *Introduction to Computational Cardiology: Mathematical Modeling and Computer Simulation*, DOI 10.1007/978-0-387-76686-7_3,

Intercalated disks connect the cells in the bundle in an axial direction. In normal conditions and assuming a macro approach, the heart muscle can be considered a continuous medium – "functional syncytium" with axial and transversal directional anisotropy.

The geometry of a cell is reduced to a cylindrical form for rough calculations. The dimensions for a mammalian idealized cardiac cell are shown in Fig. 2

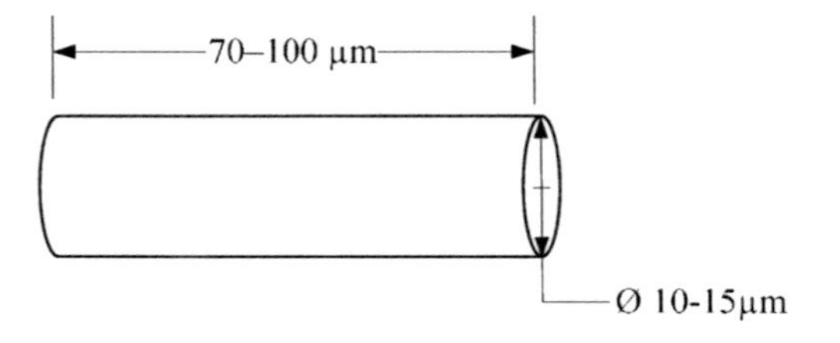

Fig.2 Cardiac cell simplified shape and dimensions

The anatomical sketch of a myocardial cell is represented schematically in Fig.3. These cells have a regular structure: the myofibrils occupy most of the cell volume and are located along the cell's longitudinal axis. The big mitochondria are stationed between them. The nucleus is localized in the center of the cell and elongated along its length.

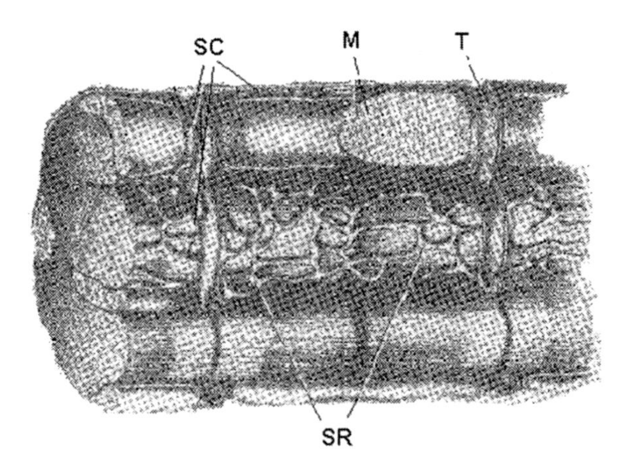

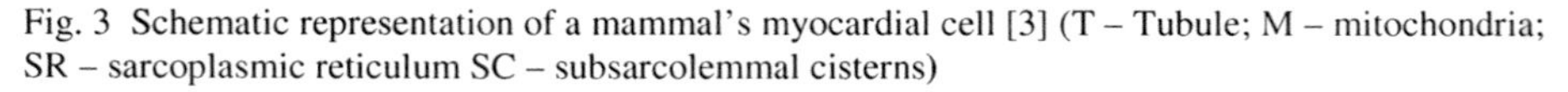

Fig. 3 Schematic representation of a mammal's myocardial cell [3] (T – Tubule; M – mitochondria; SR – sarcoplasmic reticulum SC – subsarcolemmal cisterns)

3.1.1. Intercalated disc

The intercalated disc represents one of the important structural cardiac cell systems. It provides the mechanical and electrical junctions for the neighboring cells in the axial direction and is formed by the membranes of these cells. In the axial direction of a cell, the intercalated disc has a zigzag form with the deep penetration (in a three dimensional plane) of one neighboring cell to the other (see Fig. 4). This leads to an increase in the contact surface by a factor of nine. The intercalated disc is characterized by an area with a normal intercellular membrane gap (with width on the order of 200Å) and three specialized intercellular contacts: areas of myofibrils entering a membrane (the places where myofibrils attach), desmosomes (points fastening the areas of membrane between myofibrils), and nexuses (electrical gap junctions through which the cells are connected electrically in tissue).

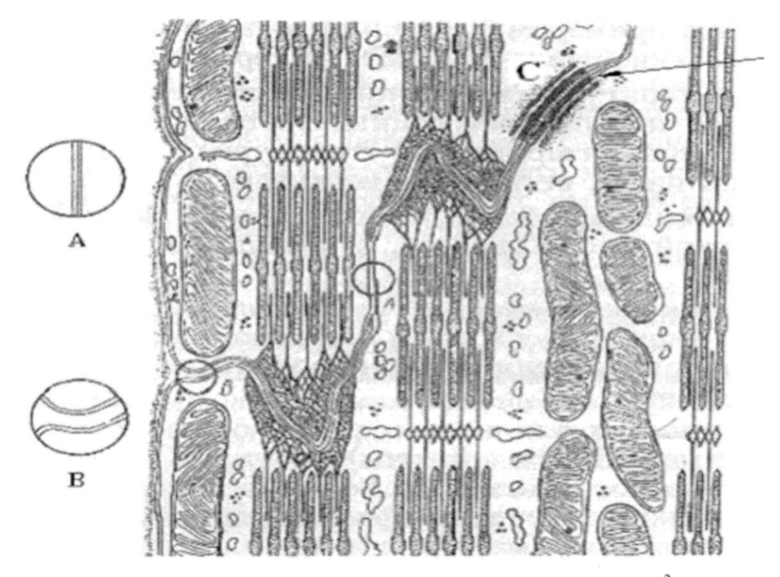

Fig. 4 Part of intercalated disc (A – Nexsus (gap junction); R_N = 1.1 Ωcm^2; B – Normal split between the neighbor membranes; C – The region of cells mechanical adhesion (Demosoma)) [5]

The side surfaces of neighboring cellular membranes are also interconnected by the specialized structures of desmosomes and nexuses. The latter are encountered rarely on the lateral side of the membranes and are more frequently found in intercalated discs. That explains the anisotropy of conductance in cardiac tissue in the axial and transversal directions (conductance in the axial direction is approximately 9 times higher than in the transversal direction).

3.1.2. Myofibrils

Myofibrils are defined as bundles of myofilaments [6]. Myofilaments are the contractile elements of cardiac cells responsible for the transformation of chemical energy into mechanical work required for heart contraction in the process of blood pumping. The myofilaments occupy 45-60% of the cell volume. They are composed of the thin and thick filaments of two major proteins, actin and myosin respectively. The thin actin filaments (~10nm thick) extend 1 μm from the Z-line toward the center of each sarcomere (see Fig. 5). The thick myosin filaments are ~1.6 μm long and 15 nm thick. Titin is a long structural protein that runs from the M-line, through the thick filament and all the way to the Z-lines. It plays a role in the structural foundation for myosin deposition on the thick filaments. The interaction between actin and myosin caused by increased $[Ca]_i$ during AP leads to the contraction of each heart muscles' cells as well as the heart as a whole. The detailed explanation of the mechanisms of heart muscle contraction processes on the molecular level can be found in Bers' excellent book [6].

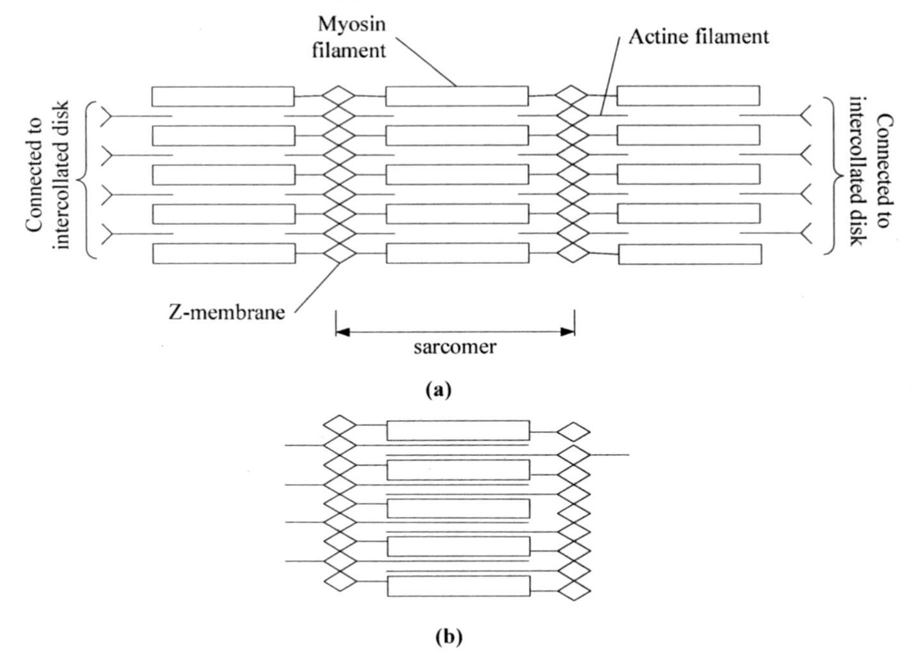

Fig. 5 Schematic representation of a myofibril. (a) Position of filaments when $[Ca^{2+}]_i$ is low; (b) Position of filaments when $[Ca^{2+}]_i$ is high.

Here we present only the static characteristic of contractile elements (see Fig. 6 and 7), which are known from physiological literature [6, 7]

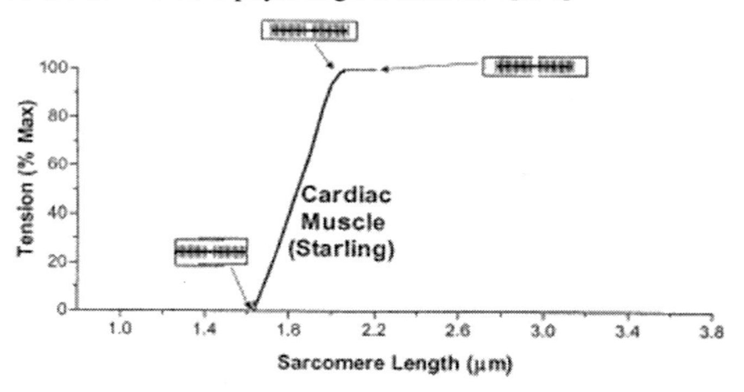

Fig. 6 The length-tension relationship for cat cardiac muscle (excerption from Fig 19 of the Bers [6] book) is presented for the range of physiological sarcomere length.

In Fig. 7 we observe the relationships between the developed cardiac muscle tension and the concentration of intracellular $[Ca^{2+}]_i$ for different temperatures for a single species and for multiple species with a single temperature. These graphs allow us to find the sensitivity of the tension to temperature and $[Ca]_i$ changes and sensitivity of tension to $[Ca^{2+}]_i$ for different species.

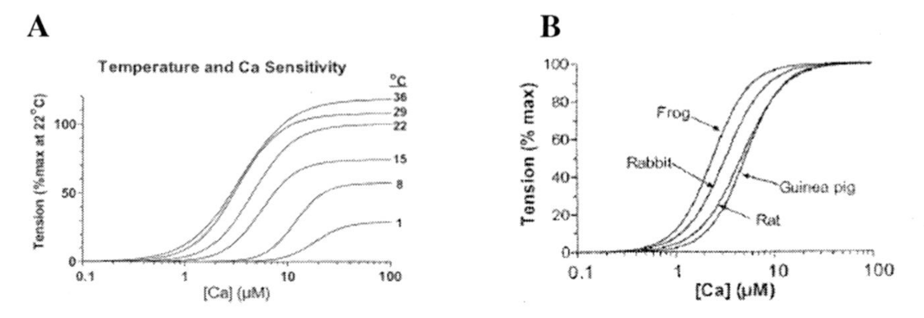

Fig. 7 A. The influence of temperature on the tension-$[Ca]_i$ relationship for chemically skinned rabbit ventricular muscle. Both the sensitivity of maximum tension to $[Ca]_i$ and temperature decrease with transfer to the lower temperatures. B. Myofilament tension-$[Ca]_i$ dependencies for ventricles of different species at temperature under 29°C (excerption from Figs. 20 and 21 of [6])

3.1.3. Cell membrane

A cell membrane forms a boundary which separates the intracellular liquid and intracellular compartments (organelles) from external liquid. The structure of a membrane is shown in Fig. 8. A membrane consists of a double layer of phospholipid molecules and aggregates of globular proteins. It also contains water filled pores and protein-lined channels.

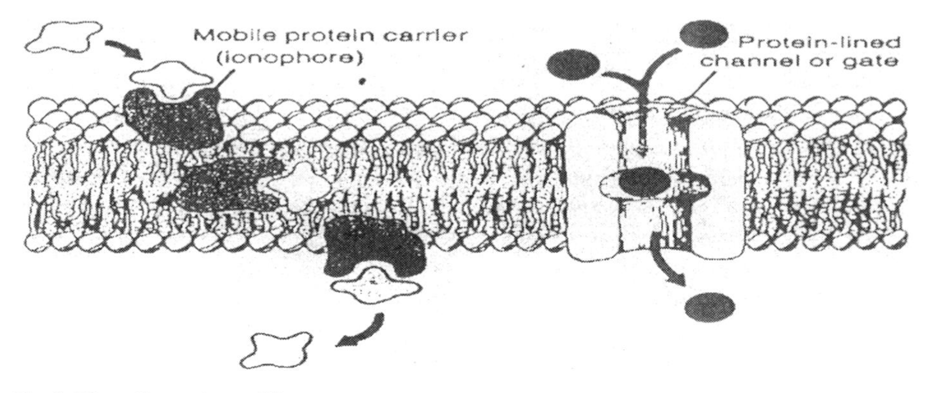

Fig. 8 The cell membrane [3]

Extra-cellular and intra-cellular liquid contain, among other things, Na^+, K^+, Cl^-, and Ca^{2+} ions in the concentrations shown in the Table 1.

Table 1. Ion concentrations for standard preparations at t = 37 ^{0}C

In extracellular liquid $[X]_e$	In intracellular liquid $[X]_i$
$[K^+]_e$ = 5.4 mM	$[K^+]_i$ = 145 mM
$[Na^+]_e$ = 140 mM	$[Na^+]_i$ = 10 mM
$[Ca^{2+}]_e$ = 1.8 mM	$[Ca^{2+}]_i$ = 0.12*10^{-3} mM
$[Cl^-]_e$ = 120 mM	$[Cl^-]_i$ = 20 mM

In an equilibrium state, the membrane has different permeability for different ions, which prevents the uniform distribution of ion concentration between extra- and intra-cellular liquids and leads to the creation of a resulting electro-chemical membrane potential (membrane rest potential). In response to pacemaker stimulus, the membrane potential changes and causes the changes in membrane permeability to different ions in such a way that AP is generated. This provides for an increase of Ca^{2+} current through the $(Ca)_L$ channels of the membrane, which causes a significant release of intracellular Ca^{2+} from the sarcoplasmic reticulum (SR) required for the contraction of myofibrils.

3.2. The basic laws and equations

A. The Nernst-Planck equation

The Nernst-Planck equation shows that the density of ion *S* flowing through a membrane, J_S is dependent on the concentration gradient of ion *S* and the gradient of potential on a membrane.

$$J_S(x) = -\frac{RT}{F} u_S \left[\frac{dC_S}{dx} + \varsigma_S C_S \frac{F}{RT} \frac{d\varphi}{dx} \right] \qquad (1)$$

Here:

R – is the universal gas constant
T – is the absolute temperature (273 ± t°C)
F – is the Faraday's constant
S – is the name of an ion
ζ_s – is a valence of s
u_S – is a mobility of "S"
C_S or [S] – is a concentration of "S"
$\varphi(x)$ - is an electrical potential at point x inside a membrane
$\varphi_i - \varphi_e = V$ – is a membrane potential
x – is a coordinate measured along the membrane thickness

B. The Nernst equation

Let us consider the particular case of (1) when $J_S(x) = 0$ and concentrations of ions [S] in extracellular liquid, $[S]_e$ and intracellular liquid, $[S]_i$ are constant. From (1) follows:

$$\frac{dC_S}{C_S} = -\varsigma_S \frac{F}{RT} d\varphi \tag{2}$$

After integration of the variables in expression (2) in the limits $(C_S)_e = [S]_e$ to $(C_S)_I = [S]_i$ and φ_e to φ_i, we finally obtain:

$$V_S = \frac{RT}{\zeta_S F} \ln \frac{[S]_e}{[S]_i}; \quad V_S = \varphi_i - \varphi_e \tag{3}$$

V_S is the equilibrium or the Nernst potential across a membrane caused by the difference in ions concentration on both sides of a membrane.

For example, let us find the Nernst potential for sodium and potassium ions separately. For t=27°C, RT/F=25.8 mV. It is known from the experimental data that $[Na^+]_e$= 140 mM, $[Na^+]_i$= 10 mM and $[K^+]_e$=5.4 mM, $[K^+]_i$=145 mM. Using (3) we obtain:

$$V_{Na} = 25.8 \ln \frac{[Na]_e}{[Na]_i} = 25.8 \ln \frac{140}{10} = 68.11 \text{ mV}$$

$$V_K = 25.8 \ln \frac{[K]_e}{[K]_i} = 25.8 \ln \frac{5.4}{145} = -85.4 \text{ mV}$$

C. Goldman-Hodgkin-Katz equation (GHK equation)

This equation is used in cases when concentrations of ions $[S]_e$ and $[S]_i$ are not constant and change over time

The assumptions used in derivation of the GHK equation are:

- The ions move across a membrane under the effects of diffusion and the electrical field in the same manner as in a free solution.
- The ions concentrations on the membrane borders are directly proportional to the ions concentrations in intra and extra cellular solutions: $(C_S)_e = \beta_S [S]_e$ and $(C_S)_i = \beta_S [S]_i$
- The dielectric constant of a membrane does not change along its thickness.
- The electrical field across a membrane is constant: $\frac{d\varphi}{dx} = const = \frac{V}{d}$
    - Here: φ – is the electric potential in a point x inside a membrane;
    - d – is the thickness of a membrane.

Let us apply these assumptions to the Nernst-Plank equation (1) for the density of S ions flux through a membrane. After integration (1) in the limits $x=0$ and $x=d$ and using the above-mentioned assumptions, we obtain:

$$J_S = \frac{\zeta_S u_S \beta_S V}{d} \frac{[S]_i - [S]_e e^{-\zeta_S V \frac{F}{RT}}}{1 - e^{-\zeta_S V \frac{F}{RT}}} \tag{4}$$

To obtain the current density through a membrane, we multiply both sides of (4) by $\zeta_S F$. Then we obtain:

$$I_s = \frac{U_s \beta_S}{d} \frac{RT}{F} \frac{\zeta_S^2 F^2}{RT} V \frac{[S]_i - [S]_e e^{-\zeta_S V \frac{F}{RT}}}{1 - e^{-\zeta_S V \frac{F}{RT}}}$$

The physical meaning of $\frac{u_S \beta_S}{d} \frac{RT}{F} = (P_S)_m$ is the maximal value of the permeability of the membrane to S ions. Generally, $(P_S) = (P_S)_m O_S$. O_S is the probability of the ion S channel to be in an open state. This probability is a function of membrane voltage and time and is determined using the clamp experiment data.

So, finally:

$$I_S = P_S \frac{\zeta_S^2 F^2}{RT} V \frac{[S]_i - [S]_e e^{-\zeta_S V \frac{F}{RT}}}{1 - e^{-\zeta_S V \frac{F}{RT}}} \tag{5}$$

This theory is called the theory of constant field due to the assumption that the gradient of the electrical field along the channel is unchanging. It also is based on the assumption that in the considered time interval the probability of crossing a membrane by "S" ions is not dependent on the presence of other ions. Therefore, this theory is used when a membrane is permeable to several kinds of ions and, particularly, in cases when the total ionic current equals zero (rest state).

For illustration, consider an example:

Let us find the rest potential for the membrane permeable to the ions of Na^+, K^+, and Cl^-.

In the rest state:

$$I_{Na^+} + I_{K^+} + I_{Cl^-} = 0$$

Using (5) for each of these currents, we obtain:

$$I_{Na^+} = P_{Na^+} \frac{\zeta_{Na}{}^2 F^2}{RT} V \frac{[Na^+]_i - [Na^+]_e\, e^{-\zeta_{Na} V \frac{F}{RT}}}{1 - e^{-\zeta_{Na} V \frac{F}{RT}}}$$

$$I_{K^+} = P_{K^+} \frac{\zeta_K^2 F^2}{RT} V \frac{[K^+]_i - [K^+]_e e^{-\zeta_k V \frac{F}{RT}}}{1 - e^{-\zeta_k V \frac{F}{RT}}}$$

$$I_{Cl} = P_{Cl}\frac{\zeta_{Cl}^2 F^2}{RT}V\frac{\left[Cl^-\right]_i - \left[Cl^-\right]_e e^{-\zeta_{Cl}V\frac{F}{RT}}}{1-e^{-\zeta_{Cl}V\frac{F}{RT}}}$$

Taking into consideration that $\zeta_{Na^+} = \zeta_{K^+} = 1$ and $\zeta_{Cl^-} = -1$ and V = V_{Rest}, we obtain the following after a simple transformation:

$$V_{rest} = -\frac{RT}{F}\ln\frac{P_{N_a^+}\left[Na^+\right]_i + P_{K^+}\left[K^+\right]_i + P_{Cl}\left[Cl^-\right]_e}{P_{N_a^+}\left[Na^+\right]_e + P_{K^+}\left[K^+\right]_e + P_{Cl}\left[Cl_i^-\right]_i} \quad (6)$$

The typical values of the resting potentials are presented in Table 2. [1]

Table 2. The values of resting potential for different cell types

Cell Type	Resting Potential (mV)
Neuron	-70
Skeletal muscle (mammalian)	-80
Skeletal muscle (frog)	-90
Cardiac muscle (atrial and ventricular)	-80
Cardiac Purkinje fiber	-90
Atria-ventricular nodal cell	-65
Sin-atrial nodal cell	-55

3.3. Currents through a cell membrane

Electrical currents have different natures and mathematical expressions. They can be divided into the following basic groups: capacitive currents, ionic currents through gated channels, ion exchangers, ATP pumps, ions leakage currents, and additional ionic currents induced by other cell ions. Different mathematical cell models have different compositions of these currents depending on the time of their formulation (reflecting the corresponding availability of experimental data) and on the specificity of the problem set to be solved using mathematical modeling and computer simulation approach. All membrane currents are divided into two groups: inward and outward. The currents, which cross a membrane from intracellular to extra-cellular domain, are called outward while those flowing in the opposite direction are called inward. In equations, inward currents are agreed to have a negative sign while outward currents have a positive sign.

3.3.1. Capacitive current

Cell membranes with extra- and intra-cellular liquids share many properties with capacitors. A membrane's double lipid layer represents the insulator while extra and intracellular liquids with different ion concentrations play a role of the capacitor's conductive plates.

$$C_m = \frac{Q}{V}$$

Q –is the electrical charge
V –is the membrane potential
From the definition of capacitance it follows:

$$C_m = \frac{k\varepsilon_0}{d} A_{cap}$$

A_{cap} – is the real membrane surface in cm^2
d – is the thickness of membrane
$k\varepsilon_0$ – is the insulator constant

Two definitions are used for membrane surface:

1. the surface calculated considering the membrane as cylinder A_{geom},
2. taking into account the real membrane configuration A_{cap}.

In mathematical modeling, membrane capacity is related to the membrane surface, expressed in cm^2. So, $C= C_m/A_{cap}$ [$\mu F / cm^2$]. For myocardial cell $A_{geom} < A_{cap}$

The capacitive current is determined as:

$$I_c = \frac{dQ}{dt} = C\frac{dV}{dt} \quad [\mu A / cm^2] \tag{7}$$

3.3.2. Ionic currents trough channels gated by membrane potential

When conditions for the existence of the equilibrium Nernst potentials are satisfied, it is possible to use the simplest formulation of the ionic channel current proposed by H-H [8]. Let us apply Ohm's law to the ionic membrane channel, for ion S:

$V = r_s I_S + V_S$
r_s – is a membrane resistance, generally a nonlinear function of two variables: V and time t.

Thus, expression for ionic current

$$I_S = \frac{1}{r_S}(V - V_S) = g_S(V,t)(V - V_S) \tag{8}$$

The equivalent circuit diagram constructed using (8) is shown in Fig. 9.

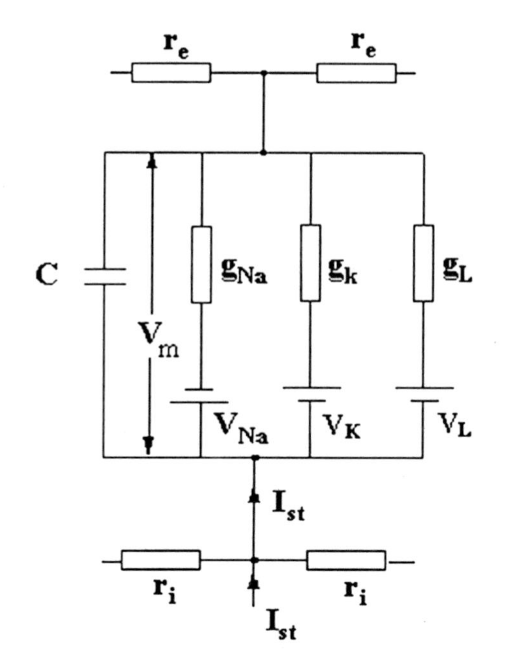

Fig. 9 The equivalent circuit diagram of a simplified cell membrane mathematical model. Three ionic currents (Sodium, Potassium and Leakage) are represented here: I_{Na}, I_K, and I_L. V_{Na}, V_K and V_L are the corresponding Nernst potentials. g_{Na}, g_K and g_L show the full conductivity of the considered ionic channels and r_i and r_e are the intra and extra cellular resistances. I_{St} represents an external stimulus current for an isolated cell or the resulting current from the neighboring cell in tissue.

The ionic current, I_S, traveling through the membrane can be expressed according to HH formulation as:

$$I_S = \overline{g}_S O_S (V - V_S),$$

Here: $\overline{g_S}$ – is the maximum value of the membrane conductance

O_S– is the probability that the S channel is in an open state. Generally O_S is a function of *V* and *t*.

The expression (8) is valid only when $[S]_e$ and $[S]_i$ are constant. In cases when $[S]_e$ and $[S]_i$ are variable, the current I_S must be computed using the GHK constant field equation:

$$I_s = (P_S)_m O_S \frac{\varsigma_S^2 F^2}{RT} V \frac{[S]_i - [S]_e e^{-\varsigma_S V \frac{F}{RT}}}{1 - e^{-\varsigma_S V \frac{F}{RT}}} \tag{9}$$

In both cases, according to H-H the function $O_S(V, t)$ is determined to be the product of the gate variables $(y_i)_S$, which represent the probabilities of possible channel states as a function of membrane potential and time. Each gate variable y_i is defined as a solution to a corresponding ordinary nonlinear differential equation of the form:

$$\frac{dy_i}{dt} = \frac{y_{i\infty}(V) - y_i}{\tau_{y_i}(V)} \tag{10}$$

Here: $y_{i\infty}$ (V) is the value of gate variable y_i at time $t = \infty$, and $\tau_{y_i}(V)$ is the time constant of this gate variable as a function of membrane voltage. The index i=(1,2,3...), reflects the possible conformation states occupied by protein molecules in ionic channel S under the effect of the membrane potential [8].

In determination of O_S, we distinguish between two approaches based on the clamp experiment data: one proposed by H-H [8] on the earlier stages of clamp experiment development (clamp on cell patches) and the other developed more recently [9] (clamping on a single ionic channel of the cell) and called, due to the random nature of channel current, the Markovian approach [10, 11].

3.4. Action potential mathematical models

The application of Kirchhoff's law to the isolated cell membrane gives:

$$C\frac{dV}{dt} + \sum_{1}^{n} I_{ion,S} + I_{st} = 0 \tag{11}$$

Here: $I_{ion,S}$ is a current of ion, S; I_{st} is a stimulus current and must have inward direction.

It is convenient to represent the total ionic current through the membrane also as a sum of currents flowing from extra-cellular into intra-cellular space and vice versa as the total inward and outward currents respectively and to assign to inward currents the minus sign. Then:

$\sum_{1}^{n} I_{ion,S} = -\sum_{1}^{l} (I_{ion,S})_{inw} + \sum_{l}^{n} (I_{ion,S})_{outw}$ and equation (11) can be rewritten as:

$$C\frac{dV}{dt} = \sum_{1}^{l} (I_{ion,S})_{inw} - \sum_{l}^{n} (I_{ion,S})_{outw} + I_{st} \tag{12}$$

From (12) follows that inward currents increase membrane potential V from the rest potential (depolarized membrane) while the outward currents decrease this potential toward the rest potential (repolarizing membrane).

With the progress being made in experimental technology more and more components of basic ionic currents have become known. Indeed, the model for a Purkinje cardiac cell (proposed by Noble in 1962 [12]) contained the following currents on the right side of eqn (12) (see Fig. 10):

$\sum (I_{ion,S})_{inw} = I_{Na} + I_{Na,b}$ and $\sum (I_{ion,S})_{otw} = I_K + I_{K_1} + I_L$

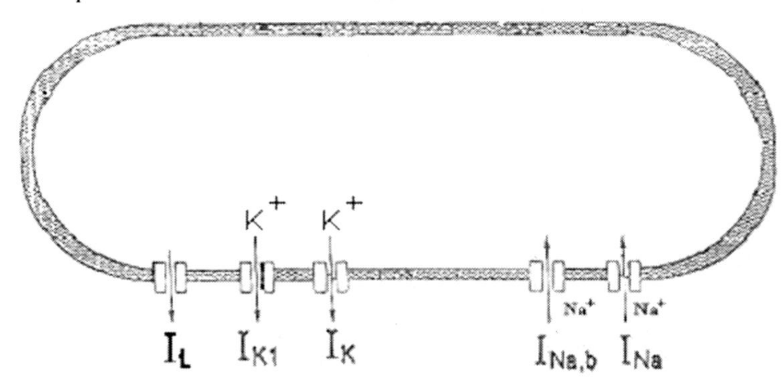

Fig. 10. Schematic diagram of the Noble cell model [12] (1962) (I_L- leakage or anion current; $I_{Na,b}$- background current)

The numbers of inward and outward currents for the model of the guinea-pig heart AP (proposed in 1994 by Luo and Rudy [13]) were significantly increased (see Fig. 11):

$$\left(\sum I_{ion,S}\right)_{inw} = I_{Na} + I_{Ca(L)} + I_{NaCa} + I_{Ca(T)} + I_{ns(Ca)} + I_{Na,b} + I_{Ca,b}$$

and

$$\sum (I_{ion,S})_{outw} = I_{K_1} + I_{K_S} + I_{K_r} + I_{K_p} + I_{NaK} + I_{p(Ca)}$$

Here, inward currents are expanded to account for Ca currents through the (L) and (T) type membrane channels. Introduction of the Na-Ca exchanger current and nonspecific, ns(Ca), currents depending on the phase of AP and intracellular Ca concentration may flow in both inward and outward directions. The outward potassium time independent current I_{K_1} is added to the plateau current I_{Kp} component and the time dependent potassium current is split into rapid I_{Kr} and slow I_{Ks} components [14]. The Ca ion pump current $I_{p(Ca)}$ is comparatively small and serves to stabilize the intracellular Ca concentration in normal conditions after cell contraction.

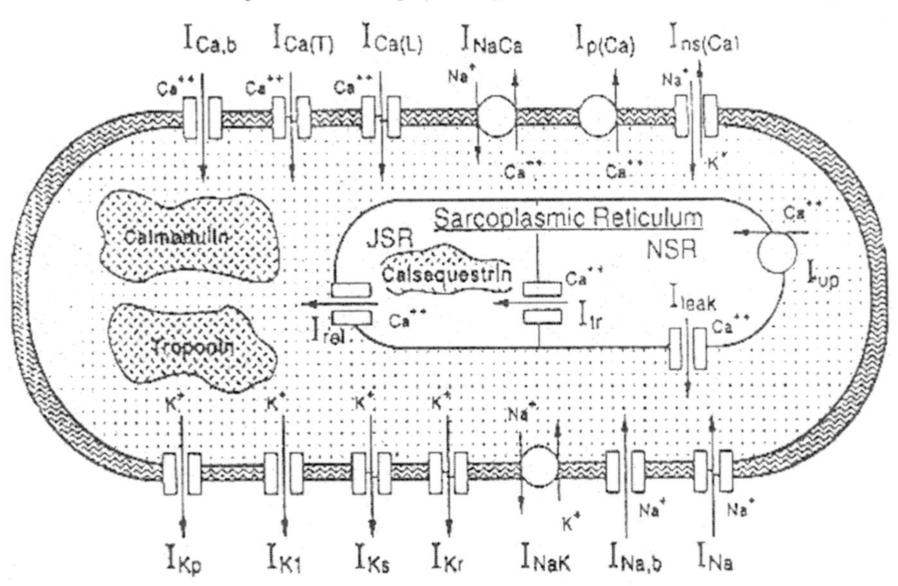

Fig. 11. Schematic diagram of the Luo-Rudy II AP model [13] (1994)

3.4.1. Action Potential and corresponding definitions.

The shape of the normal AP may be obtained by the integration of equation (11).

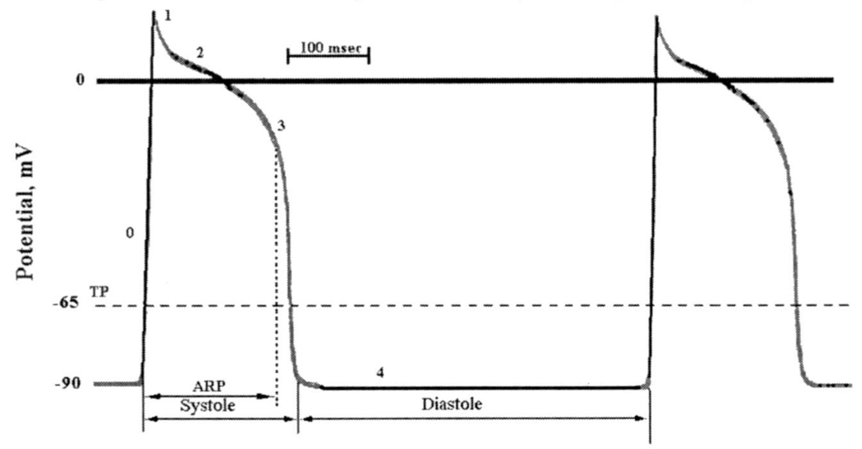

Fig.12. Example of the cardiac AP shape under a normal basic cycle length (BCL) of stimulation TP-is a threshold potential; ARP-is absolute refractory period; Systole is a total time of the AP activation and inactivation phase, Diastole is the relaxation phase of cardiac activity.

The four phases of AP are distinguished from each other as follows:

0. depolarization
1. overshoot
2. slow repolarization (plateau)

3. fast repolarization
4. equilibrium (or rest potential).

Without the application of external stimulus the cell membrane is at rest potential which is negative when the sum of ionic currents through the membrane becomes equal to zero (see eq. (6)). If this balance of inward and outward currents occurs during the repolarization phase it may cause the *prolongation of the AP*, or cardiac *cell repolarization failure.*

After the application of external stimulus as an additional inward pulse current, the membrane potential becomes linearly depolarized in time as a result of charging of the membrane capacitance. When it reaches the threshold value, the sodium channel is activated with participation of the two processes of activation (positive feedback) and inactivation (negative feedback). The interaction of these processes provides the depolarization phase-0 with prevailing inward currents. During the repolarization phases in its normal condition the cell's outward currents prevail.

The threshold potential is a membrane potential in which the balance of inward and outward currents is first distorted under the effect of the stimulus current. Its value depends on the inward and outward currents-voltage characteristics that are different for different cells and change under some pathological conditions.

The time characteristics of AP are: *AP duration,* usually measured at the 90% level from the peak of AP and basic cycle length, the time between two successive stimuli. The latter is divided into two parts: *systole* and *diastole*. During the systole, the absolute and relative refractory periods emerge. The diastole is the time when cells rest from excitation and all ionic channel gate processes return to their initial conditions.

3.4.2. Cell's passive properties

We encounter passive cell properties when membrane potential is inside the threshold range ($-V_{rest}$,$-V_{rest}$ $+V_{th}$). In this range the sum of inward and outward currents equals zero and the only inward membrane capacitance current, caused by applied pulse shape stimulus, raises the AP linearly in time. When AP reaches the upper value of the threshold ($-V+V_{th}$) the passive behavior is changed into an active one on account of prevailing inward currents (basically due to I_{Na} current). In the passive regime, total membrane conductivity is characteristically constant and equal to $g_M=\sum g_S$.

The *strength-duration relationship* is the dependence between the amplitude of the stimulus pulse $I_{st,th}$ and its duration T_{st} required to raise AP to the threshold level. To find this dependence let us consider the cell membrane equivalent circuit diagram transformed for this case, as shown in Fig. 13.

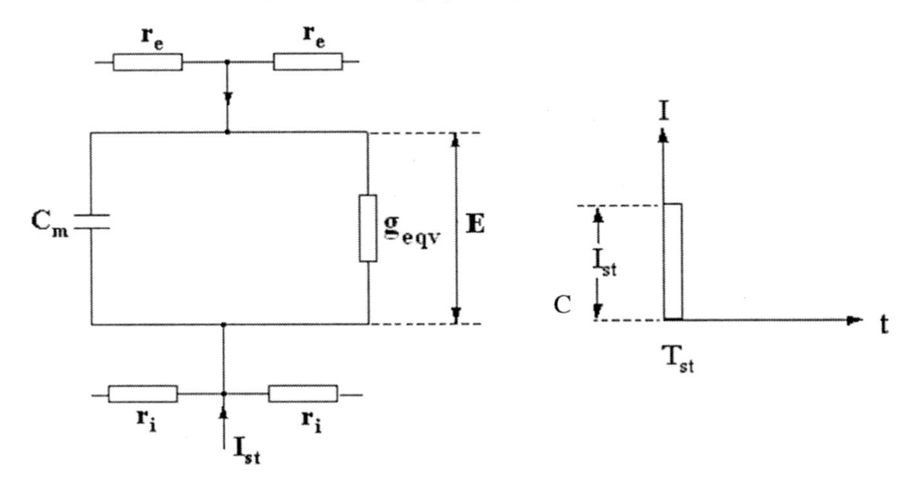

Fig. 13 Cell's equivalent circuit diagram in passive regime. For $V<V_{th}$ all g_s = const. and $\sum g_S = g_{eqv}$.

Here $E=V-V_{rest}$, so $E_{th}= V_{th}- V_{rest}$. The balance of the currents through a membrane gives:

$$C\frac{dE}{dt} + Eg_{eqv} = I_{st} \quad \text{with } E(0) = 0$$

The solution is:

$$E = \frac{I_{st}}{g_{eqv}}\left[1 - e^{\frac{-g_{eqv}t}{C}}\right] \text{ or } I_{st} = \frac{Eg_{eqv}}{1 - e^{-\frac{g_{eqv}t}{C}}} \tag{13}$$

Let us find the dependence between I_{st} and T_{st} that provides that E reaches E_{th}.
For t $=T_{st}$ and E = E_{th} we obtain:

$$I_{st} = \frac{E_{th}g_{eqv}}{1 - e^{-\frac{g_{eqv}T_{st}}{C}}}. \text{ For small } T_{st}, I_{st} \approx \frac{E_{th}C}{T_{st}} \tag{14}$$

This expression (14) is called the strength – duration relation and represents a hyperbolic curve for a fixed $E_{th}C$.

3.5. Types of Cardiac Cells

According to [2], the APs of pacemaker cells (SA node, Purkinje fibers) are shown in Figs. 14a and 14e and the APs of atrium and ventricular cells in Figs. 14.b, d correspondingly.

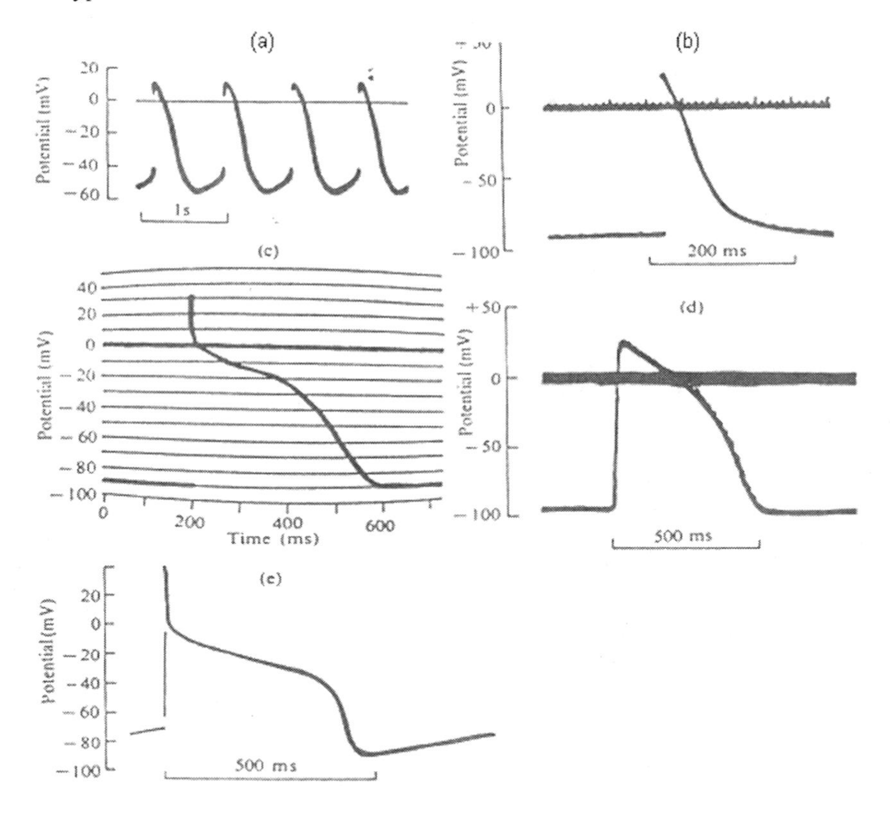

Fig. 14 Action potentials and pacemaker activity recorded in different parts of the heart.[2]: (a) Recorded from frog sinus venosus (Hutter and Trautwein 1956), (b) Recorded from dog atrium by Hoffman and Sucking (from Weidmann 1956), (c) Recorded from dog Purkinje fiber by Draper and Weidmann (1951) (photograph from Folkow and Neil (1971), (d). Recorded from frog ventricle by Hoffman (from Weidmann 1956), (e) Recorded from sheep Purkinje fiber (Weidmann 1956).

The natural pacemaker, the SA node or sinus venosus (a), is spontaneously active and the membrane potential never becomes more negative than –60 mV. Each action potential is followed by a slow spontaneous depolarization known as the pacemaker potential. The atrium (b) has a higher resting potential (which may be up to –80 or –90 mV), and a triangular-shaped action potential. It is usually quiescent, although a steady depolarizing current can induce pacemaker activity. Purkinje fibers are sometimes quiescent (c) and sometimes exhibit show pacemaker activity (e). This pacemaker activity occurs at very negative potential (-90 to –70 mV), below the range at which sinus pacemaker activity occurs. The action potential shows two phases of fast repolarization separated by a very slow phase know as the plateau. The ventricular fibers (d) have a much higher plateau and show no pacemaker activity.

The shape of APs changing across the ventricular wall. The experimental results presented in [15] are shown in Fig. 15.

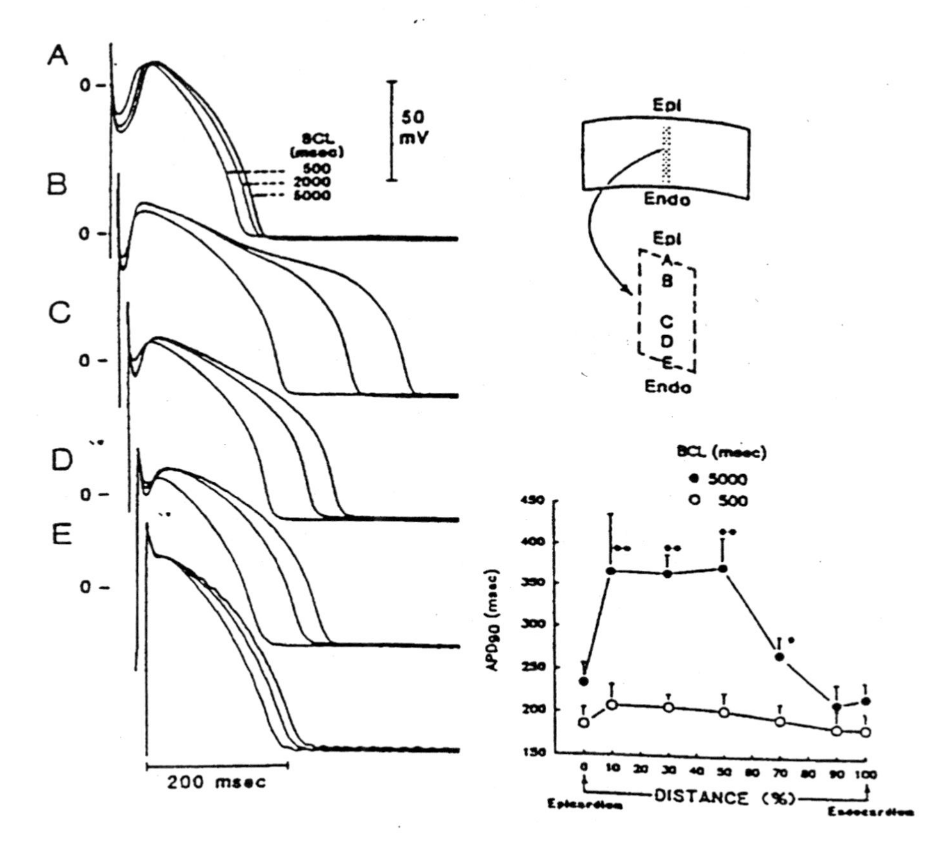

Fig. 15 The distribution of AP properties across the width of the left ventricular wall [15].

Transmembrane activity in Fig. 15 is recorded from five sites of a transmural preparation isolated from the free wall of the left ventricle near the base of the heart. Each letter in the inset indicates the respective location of the recording site. The recordings were obtained during field stimulation during the preparation. The action potentials are purposely staggered so that the upstrokes of responses from different sites do not overlap. The lower right panel shows the distribution of APD across the ventricular free wall of nine transmural preparations. APD_{90} measured at BCLs of 500 and 5000 ms is plotted as a function of the distance of the recording site of the epicardial surface (expressed as a percentage of the total width of the ventricular wall). The results suggest that M cells are widely distributed within the ventricular wall and that transitional behavior occurs from the midmyocardial to deep subendocardial layers.

3.6. References

1. Guyton, A.C. and J.E. Hall, *Textbook of Medical Physiology*. 9th ed. 1996, Philadelphia: W.B. Saunders.
2. Hoffman, B.F. and P.F. Cranefield, *Electrophysiology of the Heart*. 1960, New York: McGraw-Hill.
3. Keener, J. and J. Sneyd, *Mathematical Physiology*. 2nd ed. 2001: Springer-Verlag.
4. Zipes, D. and J. Jalife, eds. *Cardiac Electrophysiology: From Cell to Bedside*. 2nd ed. 1995, W.B. Saunders: Philadelphia.
5. Morozova, O.L., *Ultrastructure of heart cells and their junctions.* Cardiology (Russian), 1978. **12**: 121-131.
6. Bers, D.M., *Excitation-Contraction Coupling and Cardiac Contractile Force*. 2nd ed. 2001, Norwell, MA: Kluwer Academic Publishers.
7. Langer, G.A., ed. *The Myocardium*. 2nd ed. 1997, Academic Press: San Diego.
8. Hodgkin, A.L. and A.F. Huxley, *A quantitative description of membrane current and its application to conduction and excitation in nerve.* J Physiol, 1952. **117**: 500-544.
9. Sakmann, B. and E. Neher, eds. *Single Channel Recording*. 1983, Plenum Press: New York.
10. Bezanilla, F. and C.M. Armstrong, *Inactivation of the sodium channel. I. Sodium channel experiments.* J Gen Physiol, 1977. **70**: 549-566.
11. Armstrong, C.M. and F. Bezanilla, *Inactivation of the sodium channel. II. Gating current experiments.* J Gen Physiol, 1977. **70**: 567-590.
12. Noble, D., *A modification of the Hodgkin-Huxley equations applicable to Purkinje fiber action and pace-maker potentials.* J Physiol, 1962. **160**: 317-352.
13. Luo, C.H. and Y. Rudy, *A dynamic model of the cardiac ventricular action potential. I. Simulations of ionic currents and concentration changes.* Circ Res, 1994. **74**: 1071-1096.
14. Zeng, J., K.R. Laurita, D.S. Rosenbaum, and Y. Rudy, *Two components of the delayed rectifier K+ current in ventricular myocytes of the guinea pig type. Theoretical formulation and their role in repolarization.* Circ Res, 1995. **77**: 140-152.
15. Antzelevitch, C., S. Sicouri, and A. Lukas, *Clinical implications of electrical heterogeneity in the heart: The electrophysiology and pharmacology of epicardial, M and endocardial cells.*, in *Cardiac Arrhythmia: Mechanism, Diagnosis and Management*, P.J. Podrid and P.R. Kowey, Editors. 1995, William & Wilkins: Baltimore, MD. p. 88-107.

Chapter 4. Mathematical Models of Action Potential

Mathematical models of action potential first appear at the beginning and the middle of twentieth century as both models of analogy (Ostwald [1], Van Der Pol [2]), FitzHugh [3], Nagumo[4]) and pure phenomenological models (Wienner and Rosenblut [5], Moe et al [6]., Krinski [7]) including models based on finite automata representations. With significant developments of experimental technique and computer technology, and due to classical pioneering research accomplished by a group of scientists lead by Hodgkin and Huxley [8-10], the semi phenomenological ionic models have received recognition and wide applications for nerve AP models and were then modified to cardiac AP models in fundamental investigations accomplished by D. Noble and his group [11-13].

All ionic AP mathematical models are based on a balance of the electrical currents through a cell membrane. The existing ionic mathematical models reflect different knowledge of ionic currents flowing through the membrane and are based on experimental finding that ionic channel currents have stochastic character. There exist two approaches (see Chapter 3) in formulating the probability that an ionic channel, s, is in the open state. The first, introduced by Hodgkin-Huxley [8], is based on the assumption of mutual independence in time of channel gate variable processes, which describe different possible states of a channel. It is important to note that there were many concerns [14] about the validity of this assumption.

Hodjkin and Huxley acknowledged the issue of assumptive validity but maintained that, for their purposes, the issue did not affect their objective [8]:

> "...there is little hope of calculating the time course of the Sodium and Potassium conductance from first principles. Our object here is to find equations, which describe the conductances with reasonable accuracy and are sufficiently simple for theoretical calculation of <u>the action potential and refractory period</u>. For the sake of illustration we shall try to provide a physical basis for the equations, but must emphasize that the interpretation given is unlikely to provide a correct picture of the membrane."

The second approach introduces new formulation of channel gate processes based on representation of these processes as a Markov chain of interacting processes of channel state probabilities in time. The data obtained during a single channel clamp experiment [15] allows for the identification of the parameters of this formulation.

One can safely argue that H-H expressions are a particular case of Markovian representation. The AP mathematical models for cardiac cells based on Markovian representation are now in the process of intensive development.

This chapter will explore the genesis of Action Potential ionic mathematical models of cardiac cell based on H-H formulation. The latter is illustrated by description of H-H model for giant squid axonal membrane.

4.1. Hodgkin-Huxley Model for Axonal Membrane

The introduced ionic membrane currents are sodium, potassium and leakage currents:

B.Ja. Kogan, *Introduction to Computational Cardiology: Mathematical Modeling and Computer Simulation*, DOI 10.1007/978-0-387-76686-7_4,

$I_{Na} = g_{Na}(V - V_{Na})$; $I_K = g_K(V - V_K)$; and $I_l = g_l(V - V_l)$

Here: V_{Na}, V_K, V_l are the known equilibrium Nernst potentials for these ions,

I_{Na}, I_K, I_l are the known ionic currents from clamp-experiments as a family of the functions $I_{Na}(t, V_i)$, $I_K(t, V_i)$, $I_l(V_i)$ with clamp voltage V_i as a parameter

If we divide these currents respectively on their driving force $V_i - V_S$ (*S* denotes ion names, we obtain conductivities g_S:

$$g_S(t, V_i) = \frac{I_s(t, V_i)}{V_i - V_S}$$

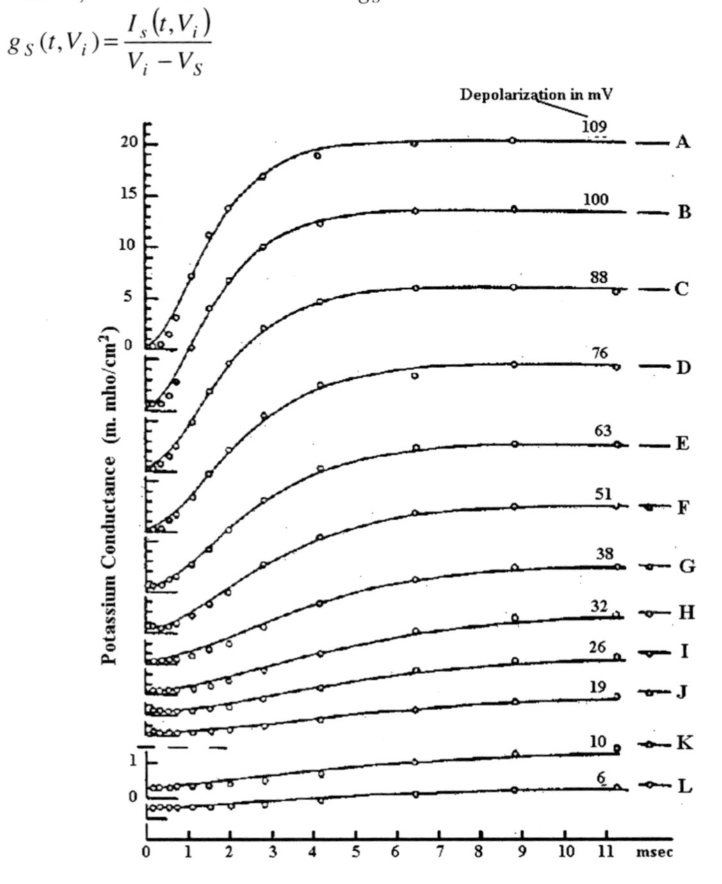

Fig. 1 Rise of potassium conductance associated with different depolarizations. The circles are experimental points obtained on axon 17, temperature 6-7°C, using observations in seawater and choline seawater [8]. The ordinate scale is the same in the upper ten curves (A to J) and is increased fourfold in the lower two curves (K and L).

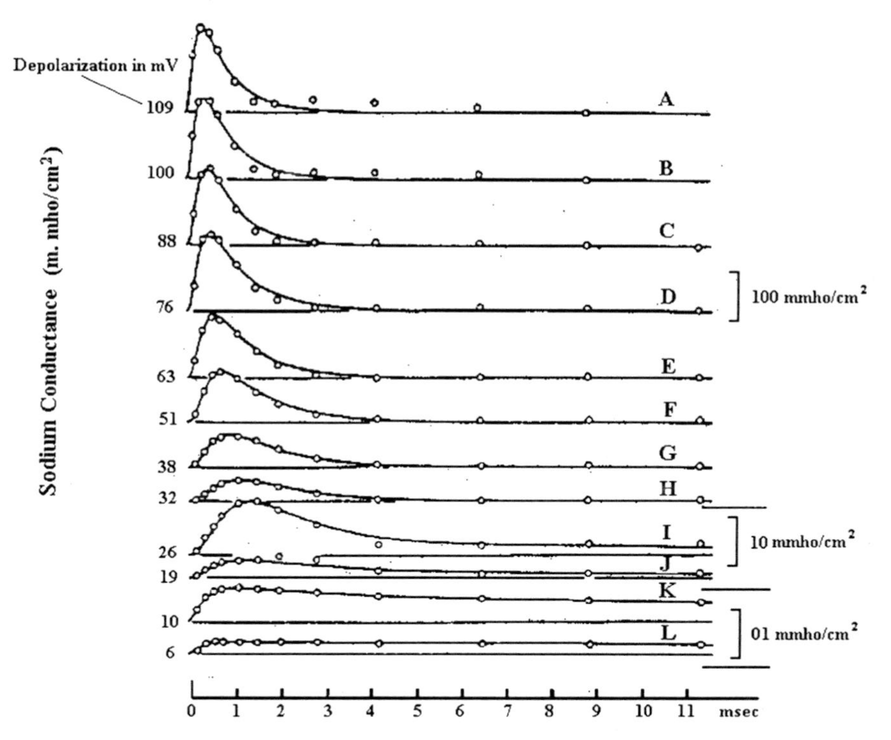

Fig. 2 Changes of sodium conductance associated with different depolarizations [8]. The circles are experimental estimates of sodium conductance obtained on axon 17, temperature 6-7°C

Families of curves for potassium and sodium conductivities are presented in figs. 1 and 2.

Analytical expressions approximating the experimental data were proposed by H-H as follows:

$$g_{Na} = \overline{g_{Na}} m^3 h \text{, } g_K = \overline{g_K} n^4 \text{, } g_l = \overline{g_l} \tag{1}$$

Here: $\overline{g_{Na}}$, $\overline{g_K}$, $\overline{g_l}$ are constant components of corresponding ion channels conductivities; *m*, *h*, and *n* are gate variables of sodium and potassium channels respectively. They are presented as continuous functions (of common time *t* and voltage *V*) obtained in the process of solving the full system of equations presented below.

The full system of equations for HH mathematical model.

The total current I_m through the membrane (ions, capacitive and stimulus I_{stim}) is:

$$I_m = C\frac{dV}{dt} + \overline{g_{Na}} m^3 h (V - V_{Na}) + \overline{g_K} n^4 (V - V_K) + \overline{g_l}(V - V_l) + I_{stim}; \tag{2}$$

The gate variables equations:

$$\frac{dm}{dt}=\alpha_m(1-m)-\beta_m m,\quad \frac{dh}{dt}=\alpha_h(1-h)-\beta_h h,\quad \frac{dn}{dt}=\alpha_n(1-n)-\beta_n n; \qquad (2a)$$

Known functions:

$$\alpha_m=0.1\frac{V-25}{1-\exp^{\frac{25-V}{10}}},\qquad \beta_m=4\exp\left(-\frac{V}{18}\right);$$

$$\alpha_h=0.07\exp\left(-\frac{V}{20}\right),\qquad \beta_h=\frac{1}{1+\exp\frac{30-V}{10}}; \qquad (2b)$$

$$\alpha_n=0.01\frac{V-10}{1-\exp\frac{10-V}{10}},\qquad \beta_n=0.125\exp\left(-\frac{V}{80}\right).$$

Parameters:

Here: $C=1\frac{\mu F}{cm^2}$, $\overline{g_{Na}}=120\,\frac{mmho}{cm^2}$, $\bar{g}_K=36\frac{mmho}{cm^2}$, $g_l=0.3\frac{mmho}{cm^2}$,

$V_{Na}=115\,mV$, $V_K=-12\,mV$, and $V_l=10\,mV$

In equations (2a and 2b), V represents the difference between membrane potential E_m and E_R rest potential. Thus $V=E_m-E_R$, $V_{Na}=E_{Na}-E_R$, $V_K=E_K-E_R$, $V_l=E_l-E_R$. For nerve cell $E_R=-60\ mV$, $E_{Na}=55\ mV$, $E_K=-72\ mV$, $E_l=-50\ mV$.

The variables m, h and n are called the gate variables. They are dimensionless, vary between 0 and 1 and control the conductivity of sodium and potassium channels respectively. The physical explanation of the sodium channel gate variables is based on the hypothesis that sodium channel opens during a temporal coincidence of three activating events (the probability of each is equal to m). The channel is closed if an inactivating event occurs with probability equal to (1-h). The gate variables m and h are assumed to change independently with time (2a).

A potassium channel is assumed to open when four activating events occur simultaneously. If the probability of one such event is n, then the probability that channel is in open state is equal to n^4.

The functions $\alpha_m(V), \beta_m(V)$; $\alpha_h(V), \beta_h(V)$ and $\alpha_n(V)$, $\beta_h(V)$ given in (2b) can be viewed as the rate constants in the first order kinetic equations for gate variables (see (2a)). These equations can be easily transformed to another form:

$$\frac{dm}{dt}=\frac{m_\infty(V)-m}{\tau_m(V)},\quad \frac{dh}{dt}=\frac{h_\infty(V)-h}{\tau_h(V)},\quad \text{and}\quad \frac{dn}{dt}=\frac{n_\infty(V)-n}{\tau_n(V)}. \qquad (3)$$

Here:

$$m_\infty(V)=\frac{\alpha_m(V)}{\alpha_m(V)+\beta_m(V)},\quad \tau_m(V)=\frac{1}{\alpha_m(V)+\beta_m(V)};$$

$$h_\infty(V)=\frac{\alpha_h(V)}{\alpha_h(V)+\beta_h(V)},\quad \tau_h(V)=\frac{1}{\alpha_h(V)+\beta_h(V)};$$

$$n_\infty(V)=\frac{\alpha_n(V)}{\alpha_{n(V)}+\beta_n(V)},\qquad \tau_n(V)=\frac{1}{\alpha_n(V)+\beta_{n(V)}}.$$

The detailed investigations of this model showed a good agreement with the major experimental data obtained for the giant squid axon. Particularly, that is true in respect to the: 1) shape, amplitude, duration of a normal AP and it threshold value; 2) shape, amplitude and conduction velocity of propagated pulse in one-dimensional nerve fiber; 3) changes in excitation threshold and response to stimulation during the refractory period.

However, like any mathematical model, which is not completely based on the first principles (semi-phenomenological) this model only represents an approximation of the reality. The authors of this model indicated that the better approximation for the potassium channel could be achieved with gate variable n raised to the fifth power, while Cole and Moore [9] suggested raising, n to the 25th power.

R. Hoyt [14, 15] considered that gate variables m and h, which determine the conductivity of sodium channel, must be presented in the model as interrelated temporal processes. These considerations acknowledge that semi-deductive mathematical models require revision each time new experimental data become available and when we try to apply them to study a new problem in the same area.

4.2. Classification of Ionic Action Potential Models

All ionic mathematical models can be provisionally divided into generations. Each generation is characterized by the time of its development, the features of the problem, which this model is assigned to solve and the relative completeness of the cardiac cell phenomena reproduced by the model.

The short characteristics of three generations of cardiac AP ionic mathematical models are presented in Fig. 3.

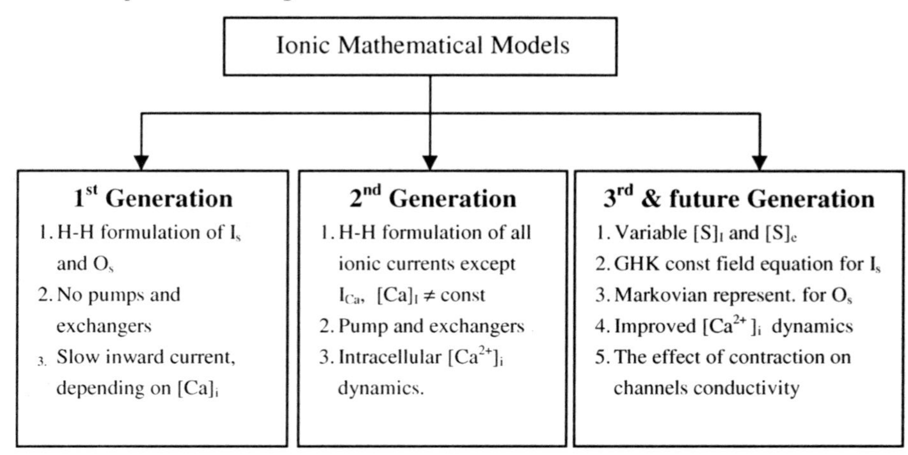

Fig. 3. Three generations of the cardiac AP ionic mathematical models

The first generation was developed during 1962-1995 and used clamp experiment data taken from both a patch of cells and a single cell. The beginning of the ***second generation models*** may be traced to 1985 and were formulated in broad outline to 1999. ***The third generation*** models are currently the subjects of intensive development.

4.3. First Generation Cardiac Action Potential Models

The evolution of the first generation of cardiac AP ionic models, including their simplified versions, are shown in Fig. 4.

These improvements pursued the major objective of extending the H-H approach proposed for the formulation of a nerve cell mathematical ionic model. The revised model was also intended to reproduce the cardiac AP and its properties in normal conditions (normal pacing rate, constant intra- and extra-cellular ionic concentrations, fixed cell volume).

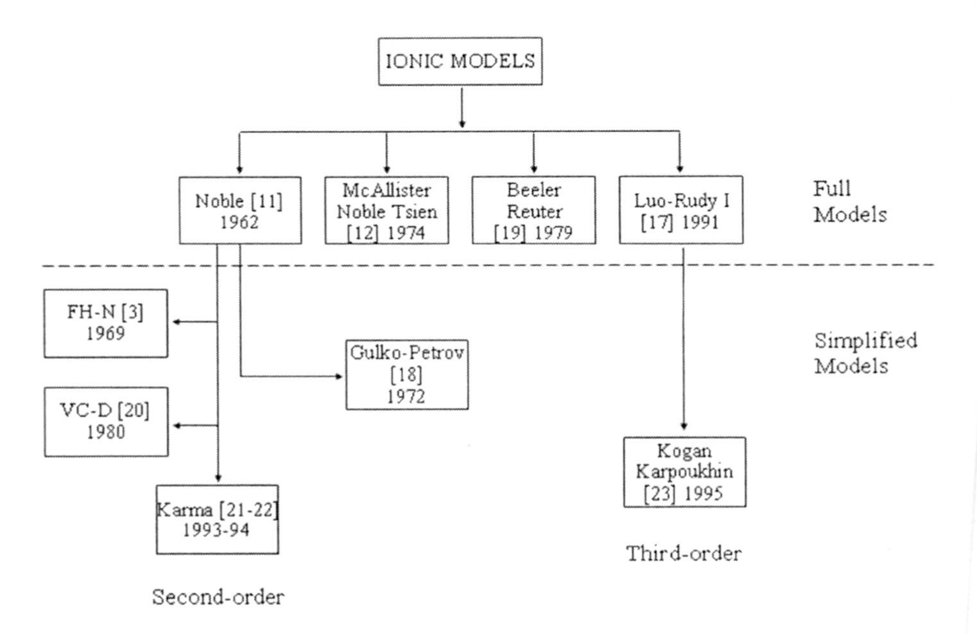

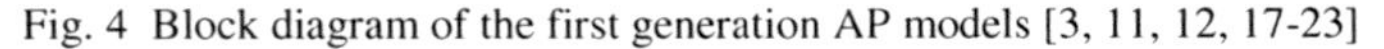
Fig. 4 Block diagram of the first generation AP models [3, 11, 12, 17-23]

The first of such cardiac AP models was proposed by D. Noble for Purkinje cardiac cells [11] and was further improved by him and other investigators mentioned in the first line in Fig. 4. The boxes shown below the dotted line relate to simplified AP models derived mostly from the cited original Noble models (for details see Chapter 5).

4.3.1. Noble Model of the Purkinje Fiber Action Potential

The H-H formulation of the AP model of the nerve cell was modified by D. Noble [11] for simulation the Purkinje cardiac cell AP. The major distinctions between the shape of a normal AP of the Purkinje fiber and that of the axon can be observed in Fig. 5.

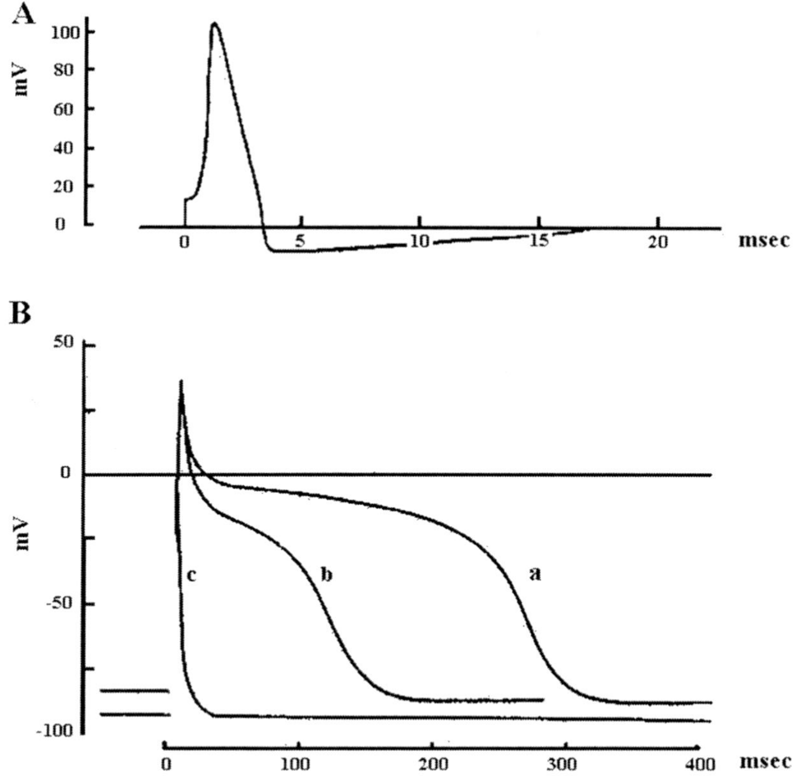

Fig. 5 Comparison of the AP for: A). Axon of the nerve cell (here E_{Rest}=–55 *mV*), and B) Purkinje fiber.

In fig. 5B is shown the effect of additional conductance, g_{K2} , on the duration of the computed AP: curve **a**, corresponds to the case without additional conductance; **b**, when additional conductance g_{K2}= 0.2 mmho/cm^2 with equilibrium potential at the resting potential; **c**, showing the effect of increasing g_{K2} by 1.0 mmho/cm^2.

Observation of Fig. 5 shows that the main distinctions between a Purkinje cardiac cell AP and a nerve cell AP are:

1. The existence of plato-phase during membrane repolarization
2. Overshoot
3. Increased duration of AP
4. Possibility to have or not to have a slow diastolic depolarization.

The equivalent electrical circuit diagrams for the Noble mathematical model of the membrane element are shown in Fig. 6.

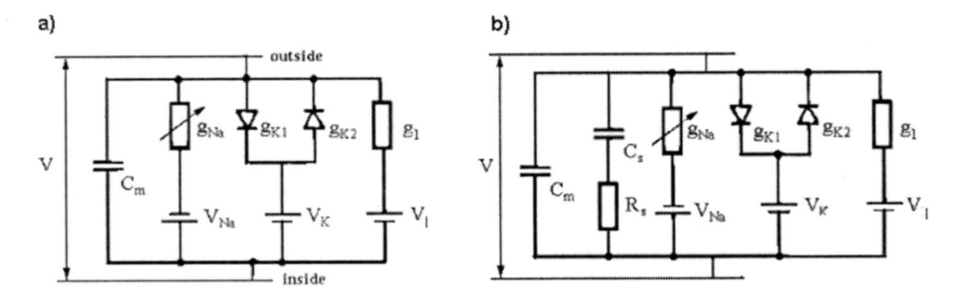

Fig. 6. Equivalent electrical circuit diagrams for a membrane element of the Purkinje fiber. a). According to Noble [11], and, b) to McAlister [12]. Here C_S is a series capacitance.

Full system of equations for Noble model has the form:

$$C\frac{dV}{dt}+g_{Na}(V-V_{Na})+g_K(V-V_K)+g_l(V-V_l)=-I_{st}; \quad (4)$$

$$g_{Na}=\overline{g_{Na_1}}\,m^3h+\overline{g_{Na_2}};\quad g_K=g_{K_1}(V)+\overline{g_{K_2}}\,n^4;$$

$$g_{K_1}=1.2\,e^{-(0.02V+1.8)}+0.015e^{(0.016V+1.5)};$$

$$\frac{dm}{dt}=\alpha_m(1-m)-\beta_m,\quad m=\frac{m_\infty(V)-m}{\tau_m(V)};$$

$$\frac{dh}{dt}=\alpha_h(1-h)-\beta_h,\quad h=\frac{h_\infty(V)-h}{\tau_h(V)};$$

$$\frac{dn}{dt}=\alpha_n(1-n)-\beta_n,\quad n=\frac{n_\infty(V)-n}{\tau_n(V)}.$$

Here:

$$\alpha_m=-\frac{0.1V+4.8)}{e^{-(0.067V+3.2)}-1},\qquad \beta_m=\frac{0.12V+0.96}{e^{(0.2V+1.6)}-1}$$

$$\alpha_h=0.17e^{-(0.05V+4.5)},\qquad \beta_h=\frac{1}{e^{-(0.1V+4.2)}+1};$$

$$\alpha_n=-\frac{0.0001(V+50)}{e^{-(0.1V+5)}-1},\qquad \beta_n=0.002e^{-(0.0125V+1.125)};$$

The constant coefficients values are:

$$\overline{g_{Na_1}}=400\,mmho/cm^2,\ \overline{g_{Na_2}}=0.14\,mmho/cm^2;$$

$$\overline{g_{K_2}}=1.2\,mmho/cm^2,\ g_l\le 0.07\,mmho/cm^2;$$

$V_{Na}=40mV$, $V_K=-100mV$, $V_l=-60mV$, and $C=12\mu F/cm^2$.

The Noble mathematical model correctly reproduces:

1. Spontaneous AP generation (pace-maker property)

2. Solitary AP generation in response to a single stimulus. The latter is achieved by increasing the $\overline{g}_{K_2}$ by 0.08 mmho/cm^2. This increase suppresses the constant component of inward current due to $\overline{g}_{Na_2}$.
3. The major characteristics of normal action potential. The exception is a maximum rate of depolarization. The later was found lowered by a factor of four to five. The addition of a series capacitor (see Fig.6b and [12]) allows the rate of depolarization to increase.

4.4. Second Generation Cardiac Action Potential Models

The models of second generation are characterized by the introduction of membrane ionic pumps, exchangers and intracellular Ca dynamics with additional $[Ca^{2+}]_i$ - sensitive currents (I_{CaL}, $I_{ns(Ca)}$, I_{NaCa}) and Ca-activated $I_{Cl(Ca)}$ currents. Additionally the potassium current in these models is represented by three components: rapid I_{Kr}, slow I_{Ks} and time independent I_{K1}.

The evolution of the second - generation AP models is shown in Fig.7 separately for the atrium and the ventricle. Basically they differ by the properties of the introduced Ca dynamic.

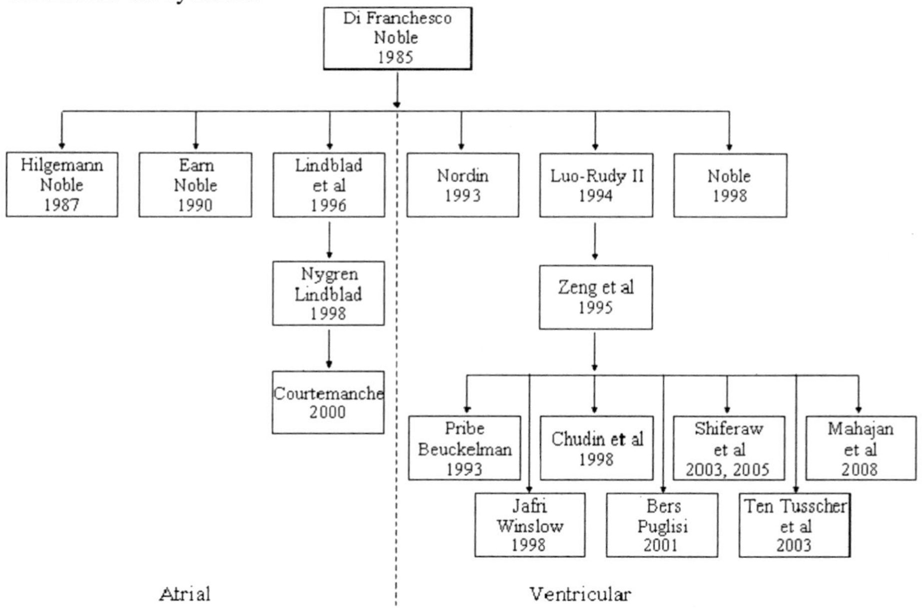

Fig. 7. Second-generation AP mathematical models

Intracellular calcium dynamics can be described as follows: during the depolarization phase of each AP a relatively small amount of Ca^{2+} enters the cell sarcoplasm via L-type Ca channels and, possibly, the reverse mode of the Na^+–Ca^{2+} exchanger. Both activate and control the release of a much larger amount of Ca^{2+}

from the sarcoplasmic reticulum (SR) known as Ca induced Ca release (CICR) process. Ca^{2+} that entered the myoplasm through the plasma membrane is pumped out of the cell primarily by a Na+- Ca^{2+}exchanger. Ca^{2+} ions that were released from the SR experience re-uptake via SR- Ca^{2+} ATPase. Other Ca^{2+} transport mechanism, such as mitochondrial Ca^{2+} uniporter and sarcolemmal Ca^{2+} pump, play a comparatively minor role under normal conditions. It should be noted, that approximately 98% of the released Ca^{2+} ions become buffered in the myoplasm. Ca^{2+} buffering in the SR increases a releasable pool of Ca^{2+} ions. Otherwise, with its small volume fraction, the SR would not be able to store sufficient amounts of Ca^{2+} ions to cause muscle contraction upon release. The processes described above are summarized in [24] (see Fig.8 below).

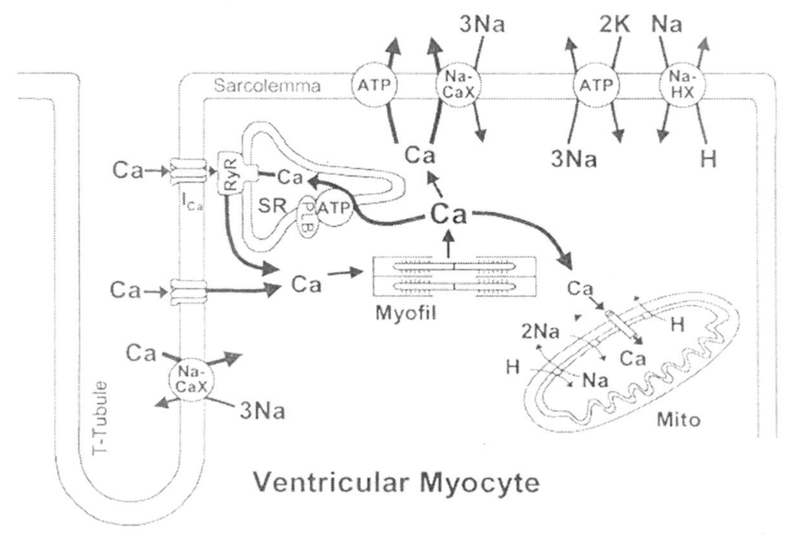

Fig. 8. General scheme of Ca cycle in a cardiac ventricular myocyte according [24]

Among the other properties of Ca dynamics, it is necessary to mention the graded response (proportionality between Ca^{2+} fluxes injected by the Ca_L channel and released with amplification (CICR) by RyRs of SR) [25], not full depletion of Ca^{2+} from JSR [26]. The latter provide under high pacing rates, the Ca^{2+} accumulation (both in sarcoplasma and JSR [24]) with following spontaneous release from SR [27]. Under high pacing rates, when CICR is formulated as a time dependent process with increased sensitivity to $[Ca^{2+}]$ concentration in JSR, the model have to reproduce Ca and APD alternans in time [28].

Intracellular Ca dynamics play a major role in the process of excitation-contractions (E-C) coupling and also produce a feedback effect on the shape and duration of AP. This feedback is mainly realized through Ca_L-dependent trans-membrane channels and Na-Ca exchanger currents. In normal conditions, the L-channel current provides the AP with no significant increased plateau region while Na-Ca exchanger prolongs the late repolarization phase. Under abnormal conditions such as brady or tachycardias complicated with LQT syndromes, the temporary distortion of repolarization currents balance (prevailing inward currents) may cause the appearance of early after depolarization (EAD) and delayed after depolarization (DAD) [29]. They often transfer heart arrhythmias into fibrillation.

The second - generation models differ mainly with respect to the completeness in reproducing the above mentioned intracellular Ca dynamics properties. Below, we discuss the handling of these properties in different proposed AP models.

4.4.1. Short Overview of Second Generation AP Models

The first AP mathematical model, incorporating intracellular Ca dynamics (DiFrancesco and Noble, [13]), was formulated using the experimental data from Purkinje fiber strands (multi-cellular preparation) and served as a prototype for subsequent models. The latter can be divided into two groups: models of the atrial myocyte AP (Hilgemann and Noble [30], Earm and Noble [31], Lindblad et al. [32], Nygren and Lindblad [33], Courtemanche [34]) and models of the ventricular myocyte AP (Nordin [35], Luo and Rudy [17], Zeng et al. [36], Priebe and Beuckelmann [37], Jafri and Winslow [38], Noble et al. [39], Chudin et al. [40], Shiferaw et al [41, 42]). Recently, new AP models were proposed for rabbit and human ventricles (see Puglisi and Bers [43], and Mahajan & Shiferaw et al [28], and K. H. W. J. ten Tusscher et al [44].

The DiFrancesco and Noble (DN) model introduced the Ica_f (an analog of the currents through Ca_L-L type channel). This current is activated by a gate variable controlled by membrane potential and inactivated by two gate variables: one, controlled by membrane potential, and the other by $[Ca^{2+}]_i$. The Na-Ca exchanger, with stoichiometry of exchange 3:1 or 4:1, is considered to be driven by Na^+ and Ca^{2+} ion gradients and by the membrane potential. The Ca^{2+} release from SR was assumed to be Ca^{2+}- induced and provided complete depletion of Ca from the release compartment. The buffering processes of Ca^{2+} ions in the myoplasm and SR were not considered.

In the Hilgemann and Noble (HN) model, the DN formulation of intracellular Ca dynamics was revised and modified, according to newly available experimental data. The activation and inactivation of Ca^{2+} channel were reformulated, resulting in faster activation and slower membrane potential dependent component of inactivation. The Ca^{2+} release from SR was modified to fit closely the results of Fabiato [29]. In addition, the membrane potential dependent activation of Ca release was incorporated. These properties were described using the Hodgkin-Huxley formulation. Ca^{2+} buffering in myoplasm was also introduced. This model was subsequently scaled down to the single-cell level by Earn and Noble, who also made other adjustments to achieve better agreement with experimental results available at that time.

Later, Lindblad et al. [32] formulated a new model for the rabbit atrial myocyte. In the intracellular Ca dynamics, they includes a description of Ca^{2+}buffers in all intracellular compartments; the CICR formulation is very similar to the one proposed by HN. In this model the inactivating effect of $[Ca^{2+}]_i$ on the current through the L-type Ca channel is not represented. This can be considered as a drawback. Courtemanthe [34] presented the most complete atrial AP model.

With respect to Ca dynamics, Nordin's model of the ventricular myocyte is based on the DN formulation with only minor changes to parameters values.

A more highly developed model, based on physiological data available in 1994-95, was presented by Luo and Rudy [45] and the Zeng et al.[36]. Rapid and slow components of the delayed rectifier K^+ currents were incorporated in that model. For the sake of brevity we will refer to this model as LR2.

4.4.2. The Luo and Rudy AP Model (LR2)

One can find the full formulation of this model with all used parameters in part I of the Appendix to this chapter. LR2 includes Ca^{2+} current through the L-type Ca channel; a Na-Ca exchanger current, buffering of the Ca^{2+}ions in the myoplasm and SR, CICR and spontaneous releases from SR. It has the advantage of having $[Ca^{2+}]_i$ dependent inactivation of the L-type channel and CICR mechanism allowing graded Ca^{2+} release. CICR is formulated using a phenomenological approach, as a process graded by a grater then threshold amount of Ca^{2+} influx into myoplasm during a time interval from the start of membrane depolarization to the moment when its maximal rate is attained. The time course of the CICR flux was reproduced by an exponential activation and inactivation processes with equal time constants $\tau_{act} = \tau_{inact} = 2ms$. Such formulation of CICR mechanism allows to us to get Ca overload conditions under high pacing rates (due to Ca accumulation) unlike other formulations in which SR is emptied completely every time CICR occurs. The introduced CICR threshold properties were shown in physiological studies to be incorrect and were removed from the new versions of this model [46, 47]. However, the artificially introduced dependence of CICR mechanism on Ca^{2+} influx (during first 2 ms after AP reaches its maximum depolarization rate) remains unchanged. In spite of the known drawbacks of the LR2 [48] model, it represent first of the few unique full ionic model of AP with Ca dynamics incorporating Ca spontaneous release. These models have served as a prototype for most of the proposed new AP models (at least in respect to sodium-potassium dynamics. It is worthwhile to mention that even today there is insufficient physiological data not only to create quantitative phenomenological model of CICR, but even qualitative phenomenological mathematical models of spontaneous release.

4.4.3. The Jafri et al AP Model

This mathematical model based on LR 2 formulation has a much more complex formulation of the L-type Ca channel. The description of this channel is characterized by Markovian approach and consists of eleven time-dependent gate states rather than two time independent states as in the LR2 model which used the H-H approach. The CICR release is formulated using the Keizer-Levine RyR adaptation model [49]. In this respect, it is necessary to mention that Gyorke & Fill found in [50] that this adaptation of the kinetics of the RyR model is too slow to account for graded CICR. Furthermore, mathematical modeling has shown [51] that if the Ca^{2+} adaptation kinetics is made much faster, it is possible to obtain graded control of CICR but at the expense of loss of model stability (spontaneous oscillations). Therefore, formulations of CICR including Ca^{2+} adaptation may not be robust.

Also, Jafri et al. introduced a "cleft space" into the model. As a result of the extremely small volume of the cleft compartment, the integration of the model equations requires very small time steps during activation of L-type channels and Ca^{2+} release from the SR. Even a fourth order Runge-Kutta adaptive time step algorithm adjusts time step to 0.0001 ms during the initial part of the AP. Such a small integration step significantly increases computational time required for simulation.

Investigation of the Jafri's et al model has shown that proposed new formulation of the current through the L-type Ca channel produces current transients similar to the one observed in the LR 2 model. The SR is emptied almost completely with each excitation as a result of the slow adaptation of RyR to the elevation of $[Ca^{2+}]$ in the cleft. With respect to the model behavior under various pacing rates, the Jafri's at al model showed a maximum peak $[Ca^{2+}]$ in the myoplasm at frequency equal to 2 Hz. This corresponds to the 500 ms basic cycle length (BCL). The maximum of the peak value of the $[Ca^{2+}]$ in the junctional SR (the release compartment) was a monotonically increasing function of the frequency of stimulation, but showed saturation starting from BCL = 250 ms. Overall, despite its computational complexity this model does not have obvious advantages over the LR 2 model in simulation of AP wave propagation in a tissue. Moreover, it produces erratic behavior unless myoplasmic $[Na^{+}]$ and $[K^{+}]$ are held constant [52].

4.4.4. The Chudin AP Model

The further modification of Ca dynamics in model of ventricle AP (see Chudin et al [40] and Appendix part II) is characterized by the extension the H-H formulation with Ca dynamics and elimination of all discontinuities presented in LR2 model. This model is one of the few that reproduce SR spontaneous release, and it is unique in its ability to vary the shape and timing of a spontaneous release by changing the corresponding model parameters. The Chudin model incorporates several properties of Ca dynamics that have been observed in physiological experiments: graded CICR by $I_{Ca,L}$, prevention of complete SR depletion during normal CICR, accumulation of Ca in the cytosol (Ca_i) and Ca in the junction SR (Ca_{jsr}) during high-frequency stimulation, and spontaneous Ca release (denoted in the model as the Ca^{2+} ion flux J_{spon}) dependent on Ca overload concentrations in both the SR and cytosol. The original model had two drawbacks. First, the time to peak of normal Ca transients is increased. Second, it is impossible to reproduce Ca and AP alternans at high pacing rate. The first drawback limits the model application to the cases of high pacing rates when time to pick values of Ca transients is naturally increased. This model has since been modified [53] to eliminate the second drawback by reformulating CICR as a time dependent process with a strong dependence on Ca_{jsr} (as was originally proposed by Shiferaw et al. [41]), while leaving unchanged the above mentioned Ca dynamics properties. The reformulation of CICR involved replacement of the original expression (5) for J_{CICR} :

$$J_{CICR} = G_{CICR} P_o P_v (\chi[Ca^{2+}]_{jsr} - [Ca^{2+}]_i) \tag{5}$$

by the time-dependent

$$\frac{dJ_{CICR}}{dt} = g_{CICR} P_o P_v \left(Q(Ca_{jsr}) - Ca_i\right) - \frac{J_{CICR}}{\tau_{CICR}} \tag{6}$$

with the initial condition $J_{CICR}(t = 0) = 0$. Here, J_{CICR} is the CICR flux from the JSR, $g_{CICR} = 2.0$ ms^{-2} is the conductance, and $\tau_{CICR} = 30$ ms is the average time constant of Ca sparks throughout the myocyte. The open probability P_o reflects the dependence of CICR on Ca entry via $I_{Ca,L}$ and the probability function P_v represents the voltage dependent nature of CICR. Expressions for both P_o and P_v can be found in the original Chudin model (see Appendix II. The term $Q(Ca_{jsr})$ is a function that reproduces the steep dependence of CICR on Ca concentration in the JSR. It is represented as the following piecewise-linear function:

$$Q(Ca_{jsr}) = \begin{cases} 0, & Ca_{jsr} < 0.5mM \\ Ca_{jsr} - 0.5, & 0.5mM \le Ca_{jsr} < 0.9mM \\ \chi(u_{rel} Ca_{jsr} + b_{rel}), & Ca_{jsr} \ge 0.9mM \end{cases} \tag{7}$$

where $u_{rel} = 11.0$ is the gain of CICR at high JSR loads, $b_{rel} = 10.0$ mM is the point of intersection between the second and third segments of $Q(Ca_{jsr})$, and χ is a parameter introduced in the original Chudin model to prevent complete Ca depletion from the SR during normal CICR. Comparison of expressions (1) and (2) allows us to consider (1) as a steady state solution of (2). The presentation of CICR as a process developed in time with time constant τ_{CICR} is a great contribution made by Shiferaw et al [42] which showed that Ca alternans may be obtained with other AP models at high pacing rate but with appropriate choice of the dynamics of the CICR process

In addition, let us show how the Chudin model allows us to reproduce different J_{spon} pulse morphologies, by varying model parameter values. The equation for J_{spon} in the Chudin model is:

$$J_{spon} = G_{spon}\, p(Ca_{jsr} - Ca_i) \tag{8}$$

$G_{spon} = 60$ ms^{-1} is the spontaneous release channel coefficient. Ca_{jsr}-Ca_i, the Ca^{2+} gradient between the JSR and cytosol, is the chemical driving force.

The gate variable p can be considered a probability of J_{spon} occurrence. It is described by a Hodgkin-Huxley type gate differential equation:

$$\frac{dp}{dt} = \frac{p_\infty - p}{\tau_p} \tag{9}$$

$$\tau_p = \tau_1 + \tau_2(1 - p_\infty) \tag{10}$$

τ_p, the time constant of variable p, effectively determines the duration of a J_{spon} event. If τ_p is increased (by increasing parameters τ_1 and τ_2), the duration of J_{spon} events also increases. p_∞, the steady state value of p, is a Hodgkin-Huxley type sigmoid function of two variables (Ca_i and Ca_{jsr}), though dependent on Ca_i and Ca_{jsr} rather than V. Bilinear approximation of this function is explicitly defined by the following equations for different regions of [Ca_i] –[Ca_{jsr}] space:

$$p_{\infty} = \begin{cases} 0, & \text{if region I} \\ \dfrac{Ca_{jsr} - K1}{K3 - K1}, & \text{if region II}(a) \\ \left(\dfrac{Ca_i - K2}{K4 - K2}\right)\left(\dfrac{Ca_{jsr} - K1}{K3 - K1}\right), & \text{if region II}(b) \\ \dfrac{Ca_i - K2}{K4 - K2}, & \text{if region II}(c) \\ 1, & \text{if region III} \end{cases} \quad (11)$$

If either Ca_{jsr} or Ca_i are below the lower thresholds set by parameters *K1* and *K2*, respectively, then $p_{\infty} = 0$ (see Appendix Fig. 1 *A*, region I). If both Ca_{jsr} and Ca_i exceed the lower thresholds *K1* and *K2*, but either fails to exceed the upper threshold set by parameters *K3* and *K4*, respectively, then p_{∞} is approximated by a linear function of Ca_{jsr} and Ca_i (see Appendix part II Fig. 1 *A*, region II). If both Ca_{jsr} and Ca_i exceed the upper thresholds *K3* and *K4*, then $p_{\infty} = 1$ (see Appendix Fig. 1 *A*, region III).

K1 and *K2* are the lower thresholds for a J_{spon} event to occur. These thresholds effectively determine the timing of the start of J_{spon} events in the AP cycle. During phase 3 and 4 of a normal AP, Ca_{jsr} is increasing and Ca_i is decreasing. Thus, to make J_{spon} start earlier in the AP cycle, *K1* (the Ca_{jsr} threshold) should be decreased and *K2* (the Ca_i threshold) should be increased. Table 1 presents the pairs of *K1* and *K2* values used to achieve J_{spon} events during diastole (temporally outside the AP) and during systole (during the AP).

Table 1. Timing of the J_{spon} upstroke in the AP cycle for different J_{spon} thresholds.

K1 (mM)	*K2* (μM)	Timing of J_{spon} upstroke
0.65	0.7	Diastole
0.58	1.12	Systole

The differences (*K3-K1*) and (*K4-K2*) effectively determine the amplitude of a J_{spon} event. If either of the differences are reduced (either by increasing the lower thresholds *K1* and *K2*, or decreasing the upper thresholds *K3* and *K4*), then the slope of the p_{∞} function and p_{∞} itself are increased, leading to higher-amplitude J_{spon} events.

A desired J_{spon} morphology can be obtained by setting the desired duration (choosing the appropriate value of τ_p) and amplitude (choosing appropriate values of *K3-K1* and *K4-K2*). Table 2 describes the J_{spon} morphology for two different sets of parameter values used in the AP model for the present study.

Table 2. J_{spon} morphology for two different sets of parameter values.

K3 – K1 (mM)	K4 - K2 (μM)	τ_p (ms)	J_{spon} morphology
0.5	0.6	$20 + 200(1-p_\infty)$	Low amplitude, long duration (Chudin)
0.0003	0.08	$20(1-p_\infty)$	High amplitude, short duration (LR2)

The first set of parameter values produces relatively low amplitude, long duration J_{spon} events (see Fig. 9(ii)) as in [40].

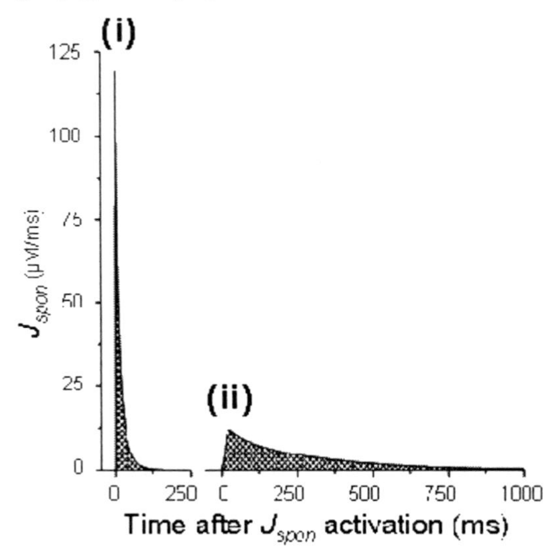

Fig. 9. Sample traces of J_{spon} activations (elicited via high-frequency stimulation) for two different sets of J_{spon} parameter values. One set (*ii*) yields low amplitude, long duration J_{spon} events. The other set (*i*) yields high amplitude (via reduction of *K3-K1* and *K4-K2*), short duration (via reduction of τ_p) J_{spon} events. The total Ca released from the SR (*crosshatched region*) is relatively equal for both morphologies.

In the second set of parameter values, τ_p is an order of magnitude smaller than in the original, while the differences (*K3-K1*) and (*K4-K2*) are at least an order of magnitude larger. Thus, it produces high amplitude, short duration J_{spon} events (see Fig. 9(*i*)), as in [45]. The total Ca in an average release was conserved between morphologies.

By first selecting the timing of the J_{spon} upstroke in the AP cycle from Table 1 (setting the values of *K1* and *K2*), and then selecting a J_{spon} morphology from Table 2 (setting the values of τ_p, *K3*, and *K4*), we obtained four cases of J_{spon} events to study: low amplitude, long duration during diastole and systole; high amplitude, short duration during diastole and systole.

4.5. Further Developments of AP Models

Noble et al. [39] proposed a AP model for guinea-pig ventricular cell, which greatly extended their previous results including accumulation and depletion of calcium in thy dyadic space (between the sarcolemma and the sarcoplasmic reticulum). The authors' strategic aim was to incorporate this model into the whole heart simulation and for the sake of computational simplicity the major Ca dynamics properties reproduced earlier in the models [38, 40, 45] were not retained. Of significant interest is the inclusion in the model of the calculation of cell contraction properties and its feedback effect on AP through additional ionic stretch channels. The authors also showed how it is possible to perform simulation study of drug receptor interaction when the drugs have kinetics with the same time scale as the cardiac AP. The proposal to use a time variable (duplicating the shape of the AP) voltage clamp during clamp experiment improves accuracy of simulation for cases when dynamics of ionic channels plays an important role (e.g. under condition of high pacing rate).

Puglisi and Bers [43] adjusted the LR2 guinea-pig AP model to a rabbit ventricle, adding in a transient outward K current (I_{t0}) and Ca-activated Cl current. The modification of the kinetics of the T-type Ca channel and the rapid component of delayed K current (I_{Kr}) were made as well as the rescaling of several conductance's to match the results in a rabbit ventricle. The authors demonstrate that by choosing the AP model parameters (corresponding to some particular case of heart failure (HF)), it is possible to decrease the level of Ca release from GSR required for delayed after depolarization (DAD) to trigger full AP. Based on this result the conclusion is made that triggered APs contribute to the nonreentrant tachycardia observed in HF.

Ten Tussher et al [44] tried to improve the previous Priebe and Beuckelmann human ventricle AP model [37] by incorporating up-to-date physiological data and some new insights on the mechanisms of AP generation. The lack of experimental information especially about Ca dynamics under the condition of fast pacing rates (for human heart cells as well as for other species) compels the authors to simplify the Ca dynamics formulation. Thus some of observed phenomena cannot be reproduced by the proposed model including Ca accumulation in sarcoplasma and JSR with following overload and JSR spontaneous release; and Ca and APD alternans occurring due to temporal characteristics of CICR (induced by I_{CaL}) from JSR. Therefore, its application to simulation of reentrant propagation in 2D heart tissue yields results as if the Ca dynamics was frozen. Never-the-less, the creation of this model may be considered a useful but premature step in an important direction.

Mahajan& Shiferaw et al [28] proposed a model of rabbit ventricular myocite AP at a rapid pacing rate. The Shannon et al model [54] (an improved version of the rabbit ventricular AP model [38]) is used here as a platform to introduce a number of important innovations. The most important are: representation CaL type channel gating as a Markov process with seven possible state including voltage and calcium induced inactivation states; and the formulation the CICR from JSR as a time dependent process, which allows reproducing Ca alternans at fast pacing rates and under some predetermined conditions. Here, as in most existing AP models (except [45] and [40]), the formulation of the spontaneous Ca release by JSR is fully absent at high pacing rates when Ca overload occurs both in the intracellular domain and in JSR. This eliminates the possibility of reproducing the DADs and EADs in a solitary cell at high pacing rates or a clusters of EADs and DADs in tissues under conditions of tachycardia and fibrillation.

In conclusion we can expect that the AP models of the third generation will express the emerging tendencies:

- to transfer from H-H formulation of channels gate processes to the Markovian representation based on single channel voltage and AP clamp experimental data;
- to introduce the effect of the variability of ionic concentrations especially under high pacing rates;
- to take into account the feedback between cell's contraction and changing of ionic channels conductivity;
- to replace the description of the CICR and GSR Ca spontaneous (for high pacing rates) releases obtained from plausible considerations and pure phenomenological basis with that based on future physiological investigations performed on a cell compartmental-molecular level.

4.6. Clamp Experiment Techniques

This technique allows time-based measurement of changes in ionic currents through the membrane when the membrane potential is fixed. In 1952 Hodgkin, Huxley and Katz made significant contributions to the development and application of the clamp-experiment method (see Fig.10A and [10]).

Most of the block-diagrams shown in Fig. 10 have common features. They use two intracellular electrodes: a voltage-recording electrode E' and a current-delivering electrode I'. The voltage electrode connects to a high impedance follower circuit (x1). The output of the follower is recorded at E and also compared with the voltage-clamp command pulses by a feedback amplifier (FBA). The highly amplified difference of these signals is applied as a current (dashed arrows) through I', across the membrane, and to the bath-grounding electrode where it can be recorded (I).

In the gap method, the extracellular compartment is divided into pools by gaps of Vaseline, sucrose, or air and the end pools contain a depolarizing "intracellular" solution. The patch-clamp method can study a minute patch of membrane sealed to the end of a glass pipette. The patch clamp represents modern improvement to the clamp-experiment technique (See Fig. 10. (E)) and allows measurement of currents through a single ionic channel [16].

A system of several intra-cellular electrodes connected to a high gain operational amplifier forms the feedback system. This system allows adjustment of the constant membrane voltage and measurement of the total ionic current and its components on a microsecond timescale. Due to the short time frame for adjustment of the membrane voltage, the total measured current does not contain a capacitive component. Further ionic separation is performed using intracellular perfusion and drugs, which selectively affect different ionic channels (see fig. 10)

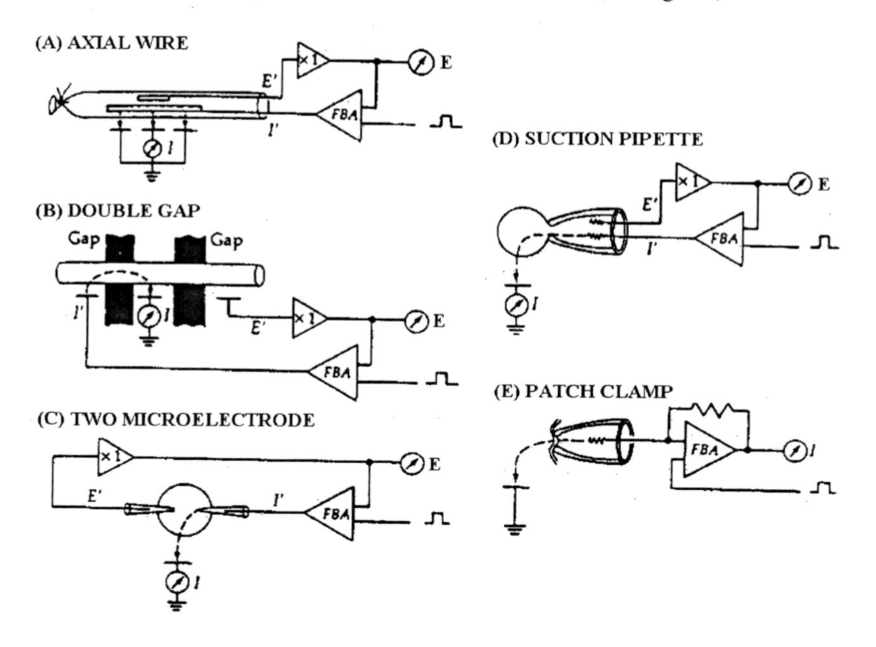

Fig. 10. Clamp-experiment techniques. (A). Using a piece of long nerve fiber membrane; (E). Patch clamp on a single cell channel

4.6.1. Separation of Na and K Currents

The classical ionic substitution method is shown in Fig. 11 following the original Hodgkin, Huxley, Katz results [10].Here ionic currents are measured in a squid axon membrane stepped from a holding potential of –65mV to –9 mV at a temperature of 8.5 °C. The components carried by Na+ ions are eliminated by substituting impairment choline ions for most of the external sodium. Algebraic difference between experimental record A and B, shows the transient inward component of current due to the inward movement of external Na^+ ions.

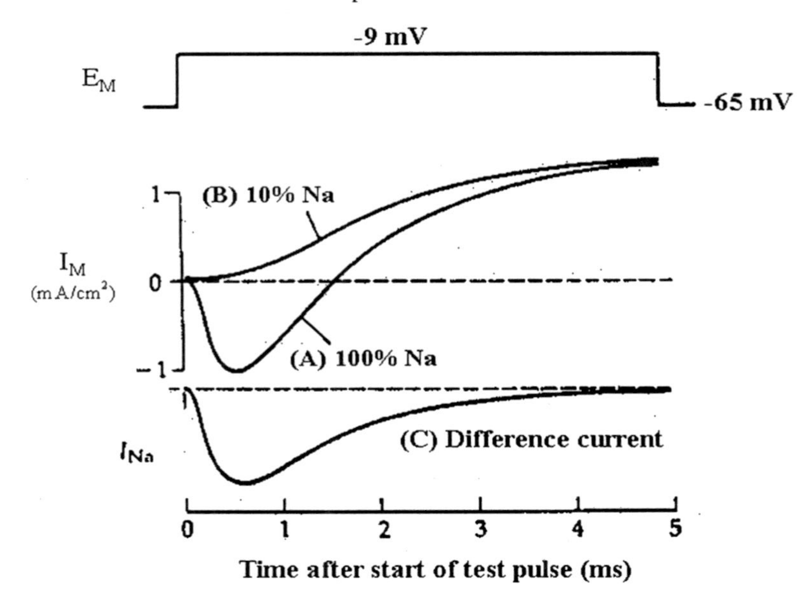

Fig.11. Separation of ionic currents using clamp-experiment data. (A) Axon in seawater, showing inward and outward ionic currents. (B) Axon in low-sodium solution with 90% of the NaCl substituted by choline chloride, showing only outward ionic current. (C).The experiments were carried out at the temperature 8.5^{0}C.

Many other drugs, channels blockers, and other substances and were found [16] for separation of ion currents in the more complex cardiac cell.

4.7. Recovery of Excitation at Normal and High Rates of Stimulation

Under normal conditions, the excitation of cardiac cells (accompanied by AP generation) occupies the systole part of the full cardiac cycle. That time span is only long enough for some state variables, such as membrane voltage and some of the channels gate variables, to return to their initial values. For illustration, all state variables are shown as functions of time for Luo and Rudy I and LR2 AP models in Fig.12 and Fig. 13.

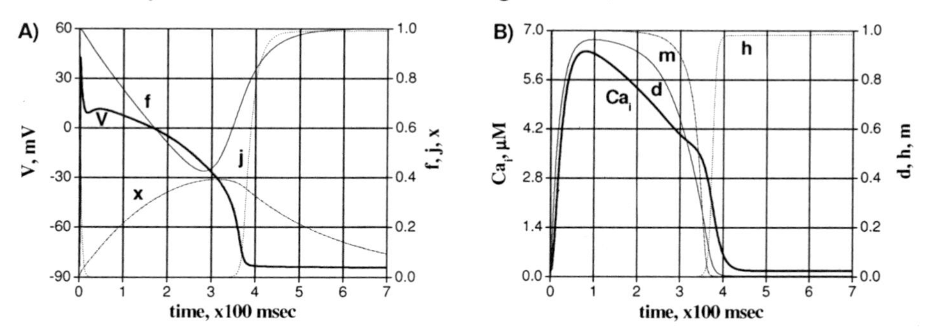

Fig. 12. A) The AP and gate variables f, j, and x; B) The gate variables d, h, m and [Ca]$_i$. in Luo-RudyI model [17].

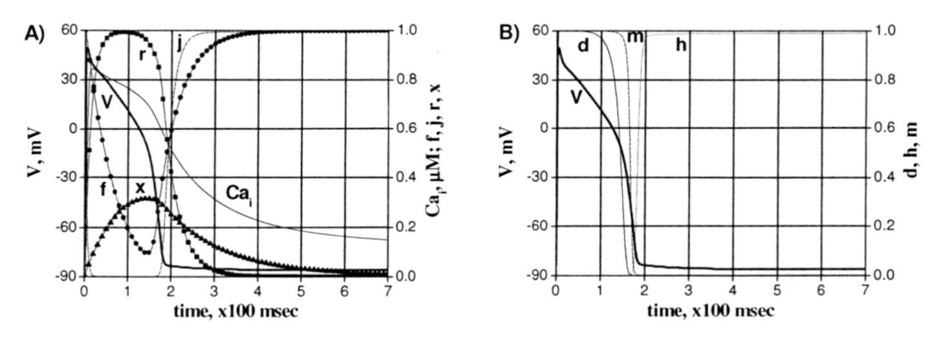

Fig. 13. A) The AP, the gate variables f, j, r, x, and [Ca]$_i$; B) The AP and gate variables d, h, and m in. Luo-Rudy II model [45].

Additional time (diastole) is required in order for the cell recovery processes to end and the next applied heart beat to produces the same AP as the previous one. The sum of this diastolic interval (DI) and systole (approximately equal to the action potential duration (APD)), composes a normal cardiac cycle.

The ability of a cardiac cell to recover after excitation is called restitution. The latter play a crucial role in cases of heart arrhythmia (brady and tachicardia) and when distortion of the balance between inward and outward currents happens during the normal cardiac cycle. The effect of cardiac cycle length shortening on the duration of APD is shown in Fig. 14 for the Nobel AP model [11].

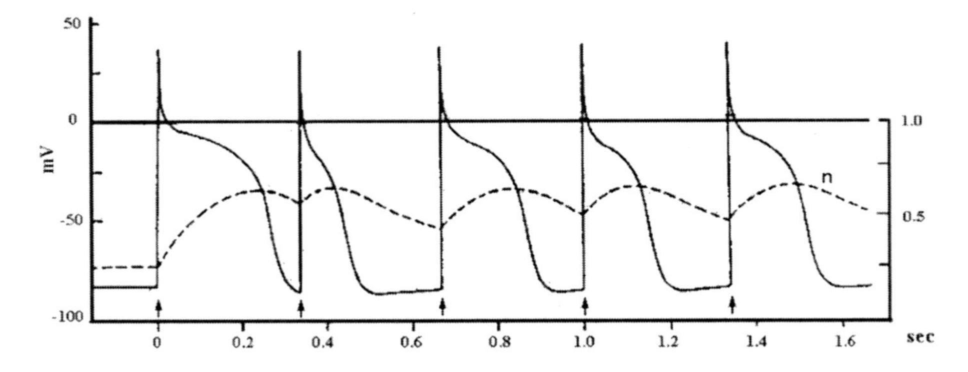

Fig. 14. The effect of repetitive stimulation on the computed action potential (continuous curve). The dotted curve shows the changes, which occur in, n, gate variable. 'Fiber made quiescent' by adding 0.1 mmho/cm^2 to g_K and then suddenly stimulated at a frequency of 3/sec. Note the alternation in duration of action potentials during transient to stationary but shorter APD.

The restitution properties of an isolated cell differ from that of the cell in tissue. Difficulties arising in experimental measurement of the recovery processes (temporal activity of membrane channel gating) attract attention to the effects, which they produce on such characteristics of AP generation and propagation as APD restitution ($APD_i = f_1\ (DI_{i-1})$ for a single cell) and conduction velocity restitution in one dimensional fibers ($CV_i = f_2(DI_{i-1})$ for some fixed position x on the fiber). Here i is the number of cardiac cycle. These restitution dependencies may be measured in physiological experiments on isolated cells or on cells located in the tissue. These measurements are much easier using computer simulations based on mathematical modeling of AP generation and propagation. In both cases, two protocols are widely used for this purpose: one is called the extra stimulus method or S_1, S_2 protocol; the other is called the dynamic method [54] or rapid pacing protocol. According to the first protocol, the myocyte is paced using a normal cardiac cycle until all measured state variable come to their stationary values. Once the stationary values are reached, an extra stimulus S_2 is delivered in progressively shorter S_1-S_2 coupling intervals. These processes continue until the loss of stimulus capture. Using the other protocol, the myocyte is paced at normal cycle length (CL) until APD reaches steady state and then CL begins to progressively decrease by some appropriately chosen time step. The myocyte is paced with new established CL for some numbers of cycles – enough to stop changes in APD. The process ends when a 2:1 block appears. The APD is measured at 90%-repolarization and the diastolic interval is calculated as CL duration minus APD_{90}.

The obtained data are fitted to the mono-exponential curve:

$$APD(T_{CLi}) = (APD)_{T_{CL\,max}}(1 - e^{-\beta T_{CLi-1}}) \tag{12}$$

Equation (12) is known as the Carmeliet approximation [56] of an APD restitution curve. Its slope, $0 \leq \gamma_{APD} \geq 1$ (derivative $d\,APD / d\,DI$ with respect to DI or CL), may serve as indicator of APD variability under different pacing rates. At the same time, these changes will cause corresponding variations in length, λ_w, of propagated wave in a tissue, which can be calculated from the self evident relationship:

$$\lambda_w = \text{APD} \times \text{CV} \tag{13}$$

The recovery processes also affect the restitution in conduction velocity (CV) for excitation propagation between cells in a tissue. CV restitution together with the APD restitution curve (when slope of both curves $\gamma > 1$ for the same DI regions) significantly increases (see (13)) the wavelength variability of propagated excitation in tissues. The latter may lead to the transfer of stationary propagated waves into nonstationary waves and even cause the wavefront to breakup (a situation resembling fibrillation).

In conclusion, it is necessary to make some comments: in reality, the APD restitution curve is not a function of the previous DI, but of all DIs encountered in the entire cell excitation history and measurements show that APD restitution curves differ between the isolated cell and those embedded in tissue. Here, it loses its single-valued properties and exhibits a loop. It is not correct to equate this restitution curve a recovery process, notwithstanding that, the restitution curves reflects these properties to some extent,. The measurements of APD and CV restitution curves in physiological experiments are not accurate and cause some difficulties.

Some researchers hold that if $\dfrac{dAPD_i}{dDI_{i-1}} \geq 1$ and $\dfrac{dCV_i}{dDI_{i-1}} \geq 1$ for some range of DI_{i-1}, the wave propagation in any tissue is unstable. The subscript i in these equations indicates the time index of the stimulation. However, this conjecture has not been proven mathematically.

4.8. Appendix: Description of AP Models with Ca^{2+} Dynamics

Part I: Luo-Rudy II Model

- **Cell geometry**

Length (L) = 100 μm; Radius (r) = 11 μm; Geometric membrane area (A_{geo}) = $2\pi r^2 + 2\pi rL = 0.767\times10^{-4}$ cm^2; Capacitive membrane area (A_{cap}) = $2A_{geo}$; Cell volume (V_{cell}) = $\pi r^2 L = 38.0\times10^{-6}$ μL; Myoplasm volume (V_{myo}) = 68%V_{cell} = 25.84×10^{-6} μL; NSR volume (V_{nsr}) = 5.52%V_{cell} = 2.098×10^{-6} μL; JSR volume (V_{jsr}) = 0.48%V_{cell} = 0.182×10^{-6} μL.

- **Standard ionic concentrations**

$[K^+]_o$ = 5.4 mM; $[K^+]_i$ = 145 mM; $[Na^+]_o$ = 140 mM; $[K^+]_i$ = 10 mM; $[Ca^{2+}]_o$ = 1.8 mM.

- **Ionic currents in the sarcolemma**

Fast sodium current: I_{Na}

$$I_{Na} = \overline{G}_{Na} m^3 hj(V - E_{Na});\ E_{Na} = \frac{RT}{F}\ln\frac{[Na^+]_o}{[Na^+]_i};\ \overline{G}_{Na} = 16\ \frac{mS}{cm^2}$$

$$\alpha_h = \alpha_j = 0;\ \beta_h = \frac{1.0}{0.13\left\{1 + \exp\left(-\frac{V + 10.66}{11.1}\right)\right\}};\ \beta_j = \frac{0.3\exp\left(-2.535x10^{-7}V\right)}{1 + \exp\left(-\frac{V + 32}{10}\right)}$$

when $V \geq -40mV$

$$\alpha_h = 0.135\exp\left(-\frac{V + 80}{6.8}\right);\ \beta_h = 3.56\exp\left(0.079V\right) + 3.1x10^5\exp\left(0.35V\right)$$

$$\alpha_j = -(V+37.78)\frac{1.2714x10^5\exp(0.2444V)+3.474x10^{-5}\exp(-0.04391V)}{1+\exp(0.311(V+79.23))}$$

$$\beta_j = \frac{0.1212\exp(-0.01052V)}{1+\exp(-0.1378(V+40.14))}$$

when $V<-40mV$

$$\alpha_m = \frac{0.32(V+47.13)}{1-\exp(-0.1(V+47.13))};\ \beta_m = 0.08\exp\left(-\frac{V}{11}\right) \qquad \forall V$$

Current through the T-type channel: $I_{Ca(T)}$

$$I_{Ca(T)} = \overline{G}_{Ca(T)} b^2 g(V-E_{Ca});\ \overline{G}_{Ca(T)} = 0.05\ \frac{mS}{cm^2};\ E_{Ca} = \frac{RT}{2F}\ln\frac{[Ca^{2+}]_o}{[Ca^{2+}]_i}$$

$$b_\infty = \frac{1}{1+\exp\left(-\frac{V+14}{10.8}\right)};\ \tau_b = 3.7 + \frac{6.1}{1+\exp\left(\frac{V+25}{4.5}\right)}$$

$$g_\infty = \frac{1}{1+\exp\left(\frac{V+60}{5.6}\right)};\ \tau_g = \begin{cases} -0.875V+12 & for\ V\leq 0\ mV \\ 12 & for\ V>0\ mV \end{cases}$$

Current through the L-type channel: $I_{Ca(L)}$

$$I_{Ca(L)} = I_{Ca,Ca}+I_{Ca,K}+I_{Ca,Na};\ For\ ion\ S = \{Ca^{2+}, K^+, Na^+\}$$

$$I_{Ca,S} = dff_{Ca}\overline{I}_S \qquad where\!: f_{Ca} = \frac{K_{m,Ca}}{K_{m,Ca}+[Ca^{2+}]_i};\ K_{m,Ca} = 0.6\ \mu M$$

$$\bar{I}_S = P_S z_S^2 \frac{VF}{RT} \frac{\gamma_{[S]_i}[S]_i \exp(\phi_S) - \gamma_{[S]_o}[S]_o}{\exp(\phi_S) - 1} \quad where\ \phi_S = \frac{z_S VF}{RT}$$

$$P_{Ca} = 5.4x10^{-4}\ \frac{cm}{\sec};\ \gamma_{[Ca]_i} = 1;\ \gamma_{[Ca]_o} = 0.341$$

$$P_K = 6.75x10^{-7}\ \frac{cm}{\sec};\ \gamma_{[K]_i} = 0.75;\ \gamma_{[K]_o} = 0.75$$

$$P_{Na} = 1.93x10^{-7}\ \frac{cm}{\sec};\ \gamma_{[Na]_i} = 0.75;\ \gamma_{[Na]_o} = 0.75$$

$$d_\infty = \frac{1}{1 + \exp\left(-\frac{V+10}{6.24}\right)};\ \tau_d = d_\infty \frac{1 - \exp\left(-\frac{V+10}{6.24}\right)}{0.035\left(V+10\right)}$$

$$f_\infty = \frac{1}{1 + \exp\left(\frac{V+35.06}{8.6}\right)} + \frac{0.6}{1 + \exp\left(\frac{50-V}{20}\right)};\ \tau_f = \frac{50}{0.985\exp\left(-0.0337(V+10)^2\right) + 1}$$

Fast component of the delayed rectifier K$^+$ current: I_{Kr}

$$I_{Kr} = \bar{G}_{Kr} x_r r_\infty (V - E_{Kr});\ \bar{G}_{Kr} = 0.02614\sqrt{\frac{[K^+]_o}{5.4}}\ \frac{mS}{cm^2},\ E_{Kr} = \frac{RT}{F}\ln\frac{[K^+]_o}{[K^+]_i}$$

$$r_\infty = \frac{1}{1 + \exp\left(\frac{V+9}{22.4}\right)};\ x_{r\infty} = \frac{1}{1 + \exp\left(-\frac{V+21.5}{7.5}\right)}$$

$$\tau_{x_r} = \frac{1}{\frac{0.00138V_1}{1 - \exp\left(-0.123V_1\right)} + \frac{0.00061V_2}{\exp\left(0.145V_2\right) - 1}};\ Where:\ V_1 = V + 14.2;\ V_2 = V + 38.9$$

Slow component of the delayed rectifier K+ current: I_{Ks}

$$I_{Ks} = \overline{G}_{Ks} x_{Ks}^2 (V - E_{Ks}); \quad E_{Ks} = \frac{RT}{F} \ln \frac{[K^+]_o + P_{Na,K}[Na^+]_o}{[K^+]_i + P_{Na,K}[Na^+]_i}; \quad P_{Na,K} = 0.01833$$

$$\overline{G}_{Ks} = 0.057 + \frac{0.19}{1 + \exp\left(\frac{pCa - 7.2}{0.6}\right)} \frac{mS}{cm^2}; \quad pCa = -\log\left([Ca^{2+}]_i\right) + 3 \text{ with } [Ca^{2+}]_i \text{ in } mM$$

$$x_{s\infty} = \frac{1}{1 + \exp\left(\frac{1.5 - V}{16.7}\right)}; \quad \tau_{x_s} = \frac{1}{\frac{7.19x10^{-5} V_1}{1 - \exp\left(-0.148 V_1\right)} + \frac{1.31x10^{-4} V_1}{\exp\left(0.0687 V_1\right) - 1}}; \quad V_1 = V + 30$$

Time-independent K+ current: I_{K1}

$$I_{K1} = \overline{G}_{K1} K1_\infty (V - E_{K1}); \quad E_{K1} = E_{Kr}; \quad \overline{G}_{K1} = 0.75 \sqrt{\frac{[K^+]_o}{5.4}} \frac{mS}{cm^2}$$

$$K1_\infty = \frac{\alpha_{K1}}{\alpha_{K1} + \beta_{K1}}; \quad \alpha_{K1} = \frac{1.02}{1 + \exp\left(0.238(V - E_{K1} - 59.215\right)}$$

$$\beta_{K1} = \frac{0l.49124 \exp\left(0.08032(V - E_{K1} + 5.476)\right) + \exp\left(0.06175(V - E_{K1} - 594.31)\right)}{1 + \exp\left(-0.5143(V - E_{K1} + 4.753)\right)}$$

Plateau K+ current: I_{Kp}

$$I_{Kp} = \overline{G}_{Kp} K_p (V - E_{Kp}); \quad \overline{G}_{Kp} = 0.00552 \frac{mS}{cm^2}; \quad E_{Kp} = E_{Kr}; \quad K_p = \frac{1}{1 + \exp\left(\frac{7.488 - V}{5.98}\right)}$$

Na^+-Ca^{2+} exchanger current: I_{NaCa}

$$I_{NaCa} = \frac{k_{NaCa}}{K_{m,Na}^3 + [Na^+]_o^3} \frac{\exp(\eta\phi)[Na^+]_i^3[Ca^{2+}]_o - \exp((\eta-1)\phi)[Na^+]_o^3[Ca^{2+}]_i}{\left(K_{m,Ca} + [Ca^{2+}]_o\right)\left(1 + k_{sat}\exp((\eta-1)\phi)\right)}$$

$$k_{NaCa} = 2000\ \frac{\mu A}{cm^2};\ K_{m,Na} = 87.5\ mM;\ K_{m,Ca} = 1.38\ mM;\ k_{sat} = 0.1;\ \eta = 0.35;\ \phi = \frac{VF}{RT}$$

Na^+-K^+ pump current: I_{NaK}

$$I_{NaK} = \frac{\bar{I}_{NaK}\bar{f}_{NaK}[K^+]_o[Na^+]_i^{1.5}}{\left(K_{m,Nai}^{1.5} + [Na^+]_i^{1.5}\right)\left([K^+]_o + K_{m,Ko}\right)};\ \bar{I}_{NaK} = 1.5\ \frac{\mu A}{cm^2};\ K_{m,Nai} = 10\ mM;\ K_{m,Ko} = 1.5\ mM$$

$$\bar{f}_{NaK} = \frac{1}{1 + 0.1245\exp\left(-\frac{\phi}{10}\right) + 0.0365\sigma\exp(-\phi)};\ \phi = \frac{VF}{RT},\ \sigma = \frac{1}{7}\left(\exp\left(\frac{[Na^+]_o}{67.3}\right) - 1\right)$$

Nonspecific Ca^{2+} activated current: $I_{ns(Ca)}$

$$I_{ns(Ca)} = I_{ns,Na} + I_{ns,K};\ \ For\ ion\ S = \left\{K^+,\ Na^+\right\}$$

$$I_{ns,S} = \frac{\bar{I}_{ns,S}[Ca^{2+}]_i^3}{K_{m,ns(Ca)}^3 + [Ca^{2+}]_i^3},\ \ K_{m,ns(Ca)} = 1.2\ \mu M$$

$\bar{I}_{ns,S}$ *is computed as* $\bar{I}_S$ *for L-channel, with* $P_{ns(Ca)} = 1.75x10^{-7}\ \frac{cm}{sec}$

Sarcolemmal Ca^{2+} pump: $I_{p(Ca)}$

$$I_{p(Ca)} = \frac{\bar{I}_{p(Ca)}[Ca^{2+}]_i}{[Ca^{2+}]_i + K_{m,p(Ca)}};\ \bar{I}_{p(Ca)} = 1.15\ \frac{\mu A}{cm^2};\ K_{m,p(Ca)} = 0.5\ \mu M$$

Ca^{2+} background current: $I_{Ca,b}$

$$I_{Ca,b} = \overline{G}_{Ca,b}(V - E_{Ca});\ \overline{G}_{Ca,b} = 0.003016\,\frac{mS}{cm^2}$$

Na^+ background current: $I_{Na,b}$

$$I_{Na,b} = \overline{G}_{Na,b}(V - E_{Na});\ \overline{G}_{Na,b} = 0.00141\,\frac{mS}{cm^2}$$

- **Ca^{2+} buffers in the myoplasm**

$$[TRP] = \overline{[TRP]}\,\frac{[Ca^{2+}]_i}{[Ca^{2+}]_i + K_{m,TRP}};\ \overline{[TRP]} = 70\ \mu M;\ K_{m,TRP} = 0.5\ \mu M$$

$$[CMD] = \overline{[CMD]}\,\frac{[Ca^{2+}]_i}{[Ca^{2+}]_i + K_{m,CMD}};\ \overline{[CMD]} = 50\ \mu M;\ K_{m,CMD} = 2.38\ \mu M$$

Here [TRP] and [CMD] are concentrations of buffered troponin and calmodulin correspondingly. The above expressions allow one to express $[Ca^{2+}]_i$ as a function of total Ca^{2+} concentration ($[Ca^{2+}]_t$) in the myoplasm:

$$[Ca^{2+}]_i = \frac{1000}{3}\left\{2\sqrt{B^2 - 3C}\,\cos\left(\frac{a\cos(E)}{3}\right) - B\right\}$$

$$B = \overline{[CMD]} + \overline{[TRP]} + K_{m,TRP} + K_{m,CMD} - 0.001[Ca^{2+}]_t$$

$$C = K_{m,TRP}\,K_{m,CMD} + \overline{[TRP]}\,K_{m,CMD} + \overline{[CMD]}\,K_{m,TRP} - 0.001[Ca^{2+}]_t\left(K_{m,TRP} + K_{m,CMD}\right)$$

$$D = -0.001[Ca^{2+}]_t\,K_{m,TRP}\,K_{m,CMD}$$

$$E = \frac{9BC - 2B^3 - 27D}{2\sqrt{\left(B^2 - 3C\right)^3}}$$

Note that both $[Ca^{2+}]_i$ and $[Ca^{2+}]_t$ are measured in μM, therefore, one needs to use a factor of 1000 to make transitions between mM and μM units.

- **Ca^{2+} fluxes in the SR**

CICR current:I_{rel}

$$I_{rel} = G_{rel}\left([Ca^{2+}]_{jsr} - [Ca^{2+}]_i\right) \frac{mM}{msec}$$

$$G_{rel} = \begin{cases} \bar{G}_{rel} \dfrac{\Delta}{K_{m,rel} + \Delta}\left[\exp\left(-\dfrac{t}{\tau}\right) - \exp\left(-\dfrac{2t}{\tau}\right)\right] & if\ \Delta > 0 \\ 0, & otherwise \end{cases}$$

$\Delta = \Delta[Ca^{2+}]_{i,2} - \Delta[Ca^{2+}]_{i,th}$; $[Ca^{2+}]_{i,th} = 0.18$ μM; $\bar{G}_{rel} = 60\ ms^{-1}$; $K_{m,rel} = 0.8$ μM; $\tau = 2$ ms

Here, $\Delta[Ca^{2+}]_{i,2}$ equals to the change in $[Ca^{2+}]_i$ 2 ms after the maximum rate of membrane depolarization has occurred; time t is reset at the beginning of each release. **Importantly, convert $[Ca^{2+}]_i$ to mM** to use the above formula.

Ca^{2+} release under Ca^{2+} overload conditions:I_{spon}

$$I_{spon} = G_{spon}\left([Ca^{2+}]_{jsr} - [Ca^{2+}]_i\right) \frac{mM}{msec}$$

$$G_{spon} = \begin{cases} \bar{G}_{spon}\left[\exp\left(-\dfrac{t}{\tau}\right) - \exp\left(-\dfrac{2t}{\tau}\right)\right] & if\ [Ca^{2+}]_{jsr} > 1.8\ mM \\ 0, & otherwise \end{cases}; \quad \bar{G}_{spon} = 4\ ms^{-1}; \ \tau = 2\ ms$$

Here, $[Ca^{2+}]_{jsr}$ is free Ca^{2+} concentration in the JSR; time t is reset with the start of each new release event. **Importantly, convert $[Ca^{2+}]_i$ to mM** to use the above formula.

Ca^{2+} buffer in the JSR

$$[CSQ] = \overline{[CSQ]}\,\frac{[Ca^{2+}]_{jsr}}{[Ca^{2+}]_{jsr} + K_{m,csq}};\quad \overline{[CSQ]} = 10\ mM;\quad K_{m,csq} = 0.8\ mM$$

Here [CSQ] is concentration of buffered calsequestrin. The above expression allows to resolve for free Ca^{2+} in the JSR ($[Ca^{2+}]_{jsr}$) as a function of total Ca^{2+} concentration ($[Ca^{2+}]_{t,jsr}$) in the JSR:

$$[Ca^{2+}]_{jsr} = \frac{1}{2}\left(b + \sqrt{b^2 + 4[Ca^{2+}]_{t,\,jsr}K_{m,csq}}\right);\quad b = [Ca^{2+}]_{t,\,jsr} - K_{m,csq} - \overline{[CSQ]}$$

Ca^{2+} uptake of NSR: I_{up}

$$I_{up} = \bar{I}_{up}\,\frac{[Ca^{2+}]_i}{[Ca^{2+}]_i + K_{m,up}}\,\frac{mM}{m\sec};\quad \bar{I}_{up} = 0.005\,\frac{mM}{m\sec};\quad K_{m,up} = 0.92\ \mu M$$

Ca^{2+} leak of NSR: I_{leak}

$$I_{leak} = \bar{I}_{leak}\,\frac{[Ca^{2+}]_{nsr}}{K_{m,nsr}};\quad K_{m,nsr} = 15\ mM;\quad \bar{I}_{leak} = 0.005\,\frac{mM}{m\sec}$$

Ca^{2+} translocation from NSR to JSR: I_{tr}

$$I_{tr} = \frac{\left([Ca^{2+}]_{nsr} - [Ca^{2+}]_{jsr}\right)}{\tau_{tr}}\,\frac{mM}{m\sec};\quad \tau_{tr} = 180\ m\sec$$

- **Membrane potential dynamic**

$$\frac{dV}{dt} = -\frac{1}{C}\left(I_{Na} + I_{Ca(T)} + I_{Ca(L)} + I_{Kr} + I_{Ks} + I_{K1} + I_{Kp} + I_{NaCa} + I_{NaK} + I_{ns(Ca)} + I_{p(Ca)} + I_{Ca,b} + I_{Na,b}\right)$$

Here C = 1 $\mu F/cm^2$ is the membrane capacitance per unit area.

• **Dynamics of gate variables**

Dynamics of all gate variables can be represented in such general form:

$$\frac{dy}{dt}=\frac{y_\infty(V)-y}{\tau_y(V)};\ \tau_y(V)=\frac{1}{\alpha_y(V)+\beta_y(V)},\ y_\infty(V)=\frac{\alpha_y(V)}{\alpha_y(V)+\beta_y(V)}$$

• **Intracellular Ca^{2+} dynamics**

$$\frac{d[Ca^{2+}]_t}{dt}=\left(I_{rel}+I_{spon}\right)\frac{V_{jsr}}{V_{myo}}-\left(I_{Ca(L)}+I_{Ca(T)}+I_{Ca,b}+I_{p(Ca)}-2I_{NaCa}\right)\frac{A_{cap}}{2V_{myo}F}-\left(I_{up}-I_{leak}\right)\frac{V_{nsr}}{V_{myo}}$$

$$\frac{d[Ca^{2+}]_{nsr}}{dt}=\left(I_{up}-I_{leak}\right)-I_{tr}\frac{V_{jsr}}{V_{nsr}}$$

$$\frac{d[Ca^{2+}]_{t,\,jsr}}{dt}=I_{tr}-I_{rel}-I_{spon}$$

Part II: Modified Model

Currents not described in this section remained the same as in the original Luo-Rudy II model given above.

Current through the T-type channel: $I_{Ca(T)}$

This current is neglected in the modified model

Na^+-Ca^{2+} exchanger current:: I_{NaCa}

$$\textit{Same as in LR II except } k_{NaCa}=1177\frac{\mu A}{cm^2}$$

Ca^{2+} uptake of NSR: I_{up}

$$I_{up}=\bar{I}_{up}\frac{[Ca^{2+}]_i^2}{[Ca^{2+}]_i^2+K_{m,up}^2}\frac{mM}{msec};\ \bar{I}_{up}=1.792x10^{-3}\frac{mM}{msec};\ K_{m,up}=0.6\ \mu M$$

Ca^{2+} leak of NSR: I_{leak}

`Importantly, convert [Ca2+]i to mM` to use the formula given below.

$$I_{leak} = g_{leak}\left([Ca^{2+}]_{nsr} - [Ca^{2+}]_i\right)\ ;\ g_{leak} = 1.195x10^{-4}\ ms^{-1}$$

Ca^{2+} translocation from NSR to JSR: I_{tr}

Same as in LR II except $\tau_{tr} = 50\ m\text{sec}$

CICR current:I_{cicr}

$$I_{cicr} = G_{cicr} P_{open} P_V\left(\chi[Ca^{2+}]_{jsr} - [Ca^{2+}]_i\right);\ G_{cicr} = 60\ ms^{-1}$$

$$P_{open} = d\,f\,f_{Ca};\ P_V = \frac{1}{1+1.65\exp(0.05V)};\ \chi = \frac{[Ca^{2+}]^3_{jsr}}{K^3_{cicr} + [Ca^{2+}]^3_{jsr}};\ K_{cicr} = 2.0\ mM$$

Here d, f, and f_{Ca} are gates of L-type Ca^{2+} channel, defined in the previous section. **Importantly, convert $[Ca^{2+}]_i$ to mM** to use the above formula.

Ca^{2+} release under Ca^{2+} overload conditions:I_{spon}

$$I_{spon} = G_{spon} P_{spon}\left([Ca^{2+}]_{jsr} - [Ca^{2+}]_i\right);\ G_{spon} = 60\ ms^{-1}$$

$$\frac{dP_{spon}}{dt} = \frac{P_\infty - P_{spon}}{\tau_p};\ \tau_p = 20 + 200\left(1 - P_\infty\right)\ m\text{sec}$$

$$P_\infty = \begin{cases} \dfrac{[Ca^{2+}]_{jsr} - K_1}{K_3 - K_1} & if\ \left([Ca^{2+}]_{jsr}, [Ca^{2+}]_i\right) \in I \\[2ex] \dfrac{[Ca^{2+}]_i - K_2}{K_4 - K_2}\dfrac{[Ca^{2+}]_{jsr} - K_1}{K_3 - K_1} & if\ \left([Ca^{2+}]_{jsr}, [Ca^{2+}]_i\right) \in III \\[2ex] \dfrac{[Ca^{2+}]_i - K_2}{K_4 - K_2} & if\ \left([Ca^{2+}]_{jsr}, [Ca^{2+}]_i\right) \in III \end{cases}$$

$$K_1 = 0.8\ mM;\ K_3 = 1.4\ mM;\ K_2 = 0.7\ \mu M;\ K_4 = 1.3\ \mu M$$

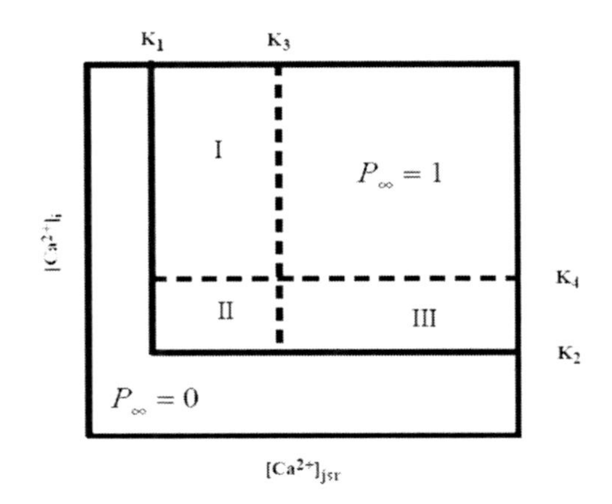

Importantly, convert $[Ca^{2+}]_i$ to mM to use the formula for I_{spon}.

Fig. 15. To the graphical representation of P_∞ as bilinear function of $[Ca^{2+}]$ and $[Ca^{2+}]$.

4.9. References

1. Ostwald, W., *Periodische Erscheinungen bei der Auflosung des Chrom in Sauren.* Zeit. Phys. Chem., 1900. 35: p. 33-76 and 204-256.
2. Van Der Pol, B. and J. Van Der Mark, *The Heartbeat Considered as Relaxation Oscillations, and an Electrical model of the Heart.* Archives Neerlandaises Physiologe De L'Homme et des Animaux, 1929. XIV: p. 418-443.
3. FitzHugh, R., *Mathematical Models of Excitation and Propagation in Nerve*, in *Biological Engineering*, H.P. Schwan, Editor. 1969, McGraw-Hill: New York. p. 1-85.
4. Nagumo, J., S. Arimoto, and S. Yoshizawa, *An Active Pulse Transmission Line Simulating Nerve Axon.* Proceedings of the IRE, 1962. 50(10): p. 2061-2070.
5. Wiener, N. and A. Rosenblueth, *The mathematical formulation of the problem of conduction of impulses in a network of connected excitable elements, specifically in cardiac muscle.* Arch. Inst. Cardiol. Mexico, 1946. 16(3): p. 205-265.
6. Moe, G.K., W.C. Rheinboldt, and J.A. Abildskov, *A computer model of atrial fibrillation.* Am Heart J, 1964. 67(2): p. 200-220.
7. Krinsky, V.I., *Fibrillation in excitable media.* Problems in Cybernetics, 1968. 20: p. 59-80.
8. Hodgkin, A.L. and A.F. Huxley, *A quantitative description of membrane current and its application to conduction and excitation in nerve.* J.Physiol., 1952. 117: p. 507-544.
9. Cole, K.S. and J.W. Moore, *Potassium ion current in the squid giant axon: dynamic characteristic.* Biophys J, 1960. 1: p. 1-14.
10. Hodgkin, A.L., A.F. Huxley, and B. Katz, *Measurement of current-voltage relations in the membrane of the giant axon of Loligo.* J Physiol, 1952. 116(4): p. 424-48.
11. Noble, D., *Modification of Hodgkin-Huxley Equations Applicable to Purkinje Fibre Action and Pace-Maker Potentials.* Journal of Physiology-London, 1962. 160(2): p. 317-&.
12. McAllister, R.E., *Computed action potentials for Purkinje fiber membranes with resistance and capacitance in series.* Biophys J, 1968. 8(8): p. 951-64.
13. DiFrancesco, D. and D. Noble, *A model of cardiac electrical activity incorporating ionic pumps and concentration changes.* Philos Trans R Soc Lond B Biol Sci, 1985. 307(1133): p. 353-98.
14. Hoyt, R.C., *The Squid Giant Axon. Mathematical Models.* Biophys J, 1963. 3: p. 399-431.
15. Hoyt, R.C., *Sodium inactivation in nerve fibers.* Biophys J, 1968. 8(10): p. 1074-97.

16. Sakmann, B. and E. Neher, eds. *Single Channel Recording*. 1983, Plenum Press: New York.
17. Luo, C.H. and Y. Rudy, *A Model of the Ventricular Cardiac Action-Potential - Depolarization, Repolarization, and Their Interaction*. Circulation Research, 1991. 68(6): p. 1501-1526.
18. Gulko, F.B. and A.A. Petrov, *On a mathematical model of excitation processes in Purkinje fiber*. Biofizika (Russian), 1970. 15(3): p. 513.
19. Beeler, G.W. and H. Reuter, *Reconstruction of the action potential of ventricular myocardial fibres*. J.Physiol.(Lond), 1977. 268: p. 177-210.
20. van Capelle, F.J. and D. Durrer, *Computer simulation of arrhythmias in a network of coupled excitable elements*. Circ Res, 1980. 47(3): p. 454-66.
21. Karma, A., *Spiral Breakup in Model-Equations of Action-Potential Propagation in Cardiac Tissue*. Physical Review Letters, 1993. 71(7): p. 1103-1106.
22. Karma, A., *Electrical alternans and spiral wave breakup in cardiac tissue*. Chaos, 1994. 4(3): p. 461-472.
23. Kogan, B.Y., W.J. Karplus, and M.G. Karpoukhin, *The third-order action potential model for computer simulation of electrical wave propagation in cardiac tissue.*, in *Computer Simulations in Biomedicine*, H. Power and R.T. Hart, Editors. 1995, Computational Mechanics Publishers: Boston.
24. Bers, D., *Excitation-Contraction Coupling and Cardiac Contractile Force*. 2nd ed. 2001, Boston: Kluwer Academic Publishers. 294-300.
25. Lopez-Lopez, J.R., P.S. Shacklock, C.W. Balke, and W.G. Wier, *Local calcium transients triggered by single L-type calcium channel currents in cardiac cells*. Science, 1995. 268(5213): p. 1042-5.
26. Bassani, J.W., W. Yuan, and D.M. Bers, *Fractional SR Ca release is regulated by trigger Ca and SR Ca content in cardiac myocytes*. Am J Physiol Cell Physiol, 1995. 268(5): p. C1313-1319.
27. Stern, M.D., M.C. Capogrossi, and E.G. Lakatta, *Spontaneous calcium release from the sarcoplasmic reticulum in myocardial cells: mechanisms and consequences*. Cell Calcium, 1988. 9(5-6): p. 247-56.
28. Mahajan, A., Y. Shiferaw, X. Lai-Hua, R. Olcese, A. Baher, M.-J. Yang, A. Karma, P.-S. Chen, A. Garfinkel, Z. Qu, and J. Weiss, *A rabbit ventricular action potential model replicating cardiac dynamics at rapid heart rates*. preprint, 2006.
29. Fabiato, A., *Time and calcium dependence of activation and inactivation of calcium-induced release of calcium from the sarcoplasmic reticulum of a skinned canine cardiac Purkinje cell*. J Gen Physiol, 1985. 85(2): p. 247-89.
30. Hilgemann, D.W. and D. Noble, *Excitation-contraction coupling and intracellular calcium transients in rabbit atrium: reconstruction of basic cellular mechanisms*. Proc.R.Soc.Lond., 1987. 230: p. 163-205.
31. Earm, Y.E. and D. Noble, *A model of the single atrial cell: relation between calcium current and calcium release*. Proc R Soc Lond B Biol Sci, 1990. 240(1297): p. 83-96.
32. Lindblad, D.S., C.R. Murphey, J.W. Clark, and W.R. Giles, *A model of the action potential and underlying membrane currents in a atrial cell*. Am.J.Physiol., 1996. 271: p. H1666-H1696.
33. Nygren, A., C. Fiset, L. Firek, J.W. Clark, D.S. Lindblad, R.B. Clark, and W.R. Giles, *Mathematical model of an adult human atrial cell: The role of K+ currents in repolariztion*. Circ.Res., 1998. 82: p. 63-81.
34. Courtemanche, M., R.J. Ramirez, and S. Nattel, *Ionic mechanisms underlying human atrial action potential properties: insights from a mathematical model*. Am J Physiol, 1998. 275(1 Pt 2): p. H301-21.
35. Nordin, C., *Computer Model of Membrane Current and Intracellular Ca2+ Flux in the Isolated Guinea Pig Ventricular Myocyte*. Am.J.Physiol., 1993. 265: p. H2117-H2136.
36. Zeng, J., K.R. Laurita, D.S. Rosenbaum, and Y. Rudy, *Two components of the delayed rectifier K+ current in ventricular myocytes of the guinea pig type. Theoretical formulation and their role in repolarization*. Circ Res, 1995. 77(1): p. 140-52.
37. Priebe, L. and D.J. Beuckelmann, *Simulation study of cellular electric properties in heart failure*. Circ Res, 1998. 82(11): p. 1206-23.

38. Jafri, M.S., J.J. Rice, and R.L. Winslow, *Cardiac Ca2+ dynamics: the roles of ryanodine receptor adaptation and sarcoplasmic reticulum load.* Biophys J, 1998. 74(3): p. 1149-68.
39. Noble, D., A. Varghese, P. Kohl, and P. Noble, *Improved guinea-pig ventricular cell model incorporating a diadic space, IKr and IKs, and length- and tension-dependent processes.* Can J Cardiol, 1998. 14(1): p. 123-34.
40. Chudin, E., J. Goldhaber, A. Garfinkel, J. Weiss, and B. Kogan, *Intracellular Ca(2+) dynamics and the stability of ventricular tachycardia.* Biophys J, 1999. 77(6): p. 2930-41.
41. Shiferaw, Y., M.A. Watanabe, A. Garfinkel, J.N. Weiss, and A. Karma, *Model of intracellular calcium cycling in ventricular myocytes.* Biophys J, 2003. 85(6): p. 3666-86.
42. Shiferaw, Y., D. Sato, and A. Karma, *Coupled dynamics of voltage and calcium in paced cardiac cells.* Physical Review E (Statistical, Nonlinear, and Soft Matter Physics), 2005. 71(2): p. 021903.
43. Puglisi, J.L. and D.M. Bers, *LabHEART: an interactive computer model of rabbit ventricular myocyte ion channels and Ca transport.* Am J Physiol Cell Physiol, 2001. 281(6): p. C2049-60.
44. ten Tusscher, K.H., D. Noble, P.J. Noble, and A.V. Panfilov, *A model for human ventricular tissue.* Am J Physiol Heart Circ Physiol, 2004. 286(4): p. H1573-89.
45. Luo, C.H. and Y. Rudy, *A dynamic model of the cardiac ventricular action potential. I. Simulations of ionic currents and concentration changes.* Circ Res, 1994. 74(6): p. 1071-96.
46. Shaw, R.M., *Theoretical studies in cardiac electrophysiology: role of membrane and gap-junctions in excitability and conduction*, in *Dept. of Biomed. Eng.* 1966, Case Western Research University.
47. Faber, G.M. and Y. Rudy, *Action potential and contractility changes in [Na(+)](i) overloaded cardiac myocytes: a simulation study.* Biophys J, 2000. 78(5): p. 2392-404.
48. Chudin, E., A. Garfinkel, J. Weiss, W. Karplus, and B. Kogan, *Wave propagation in cardiac tissue and effects of intracellular calcium dynamics (computer simulation study).* Prog Biophys Mol Biol, 1998. 69(2-3): p. 225-36.
49. Keizer, J. and L. Levine, *Ryanodine receptor adaptation and Ca2+(-)induced Ca2+ release-dependent Ca2+ oscillations.* Biophys J, 1996. 71(6): p. 3477-87.
50. Gyorke, S. and M. Fill, *Ryanodine receptor adaptation - control mechanism of Ca-induced Ca release in heart.* Science, 1993. 260: p. 807-809.
51. Stern, M.D., *Theory of excitation-contraction coupling in cardiac muscle.* Biophys.J., 1992. 63: p. 497-517.
52. Chernyavskiy, S., *Computer simulation of the Jafri-Winslow action potential model.* 1998, UCLA.
53. Samade, R. and B. Kogan. *Calcium alternans in cardiac cell mathematical models.* in *International Conference on Bioinformatics and Computational Biology.* 2007. Las Vegas, NV: CSREA Press.
54. Shannon, T.R., F. Wang, J. Puglisi, C. Weber, and D.M. Bers, *A Mathematical Treatment of Integrated Ca Dynamics within the Ventricular Myocyte.* Biophys. J., 2004. 87(5): p. 3351-3371.
55. Koller, M.L., M.L. Riccio, and R.F. Gilmour, Jr., *Dynamic restitution of action potential duration during electrical alternans and ventricular fibrillation.* Am J Physiol, 1998. 275(5 Pt 2): p. H1635-42.
56. Carmeliet, E., *Repolarisation and frequency in cardiac cells.* J Physiol (Paris), 1977. 73(7): p. 903-23.

Chapter 5. Simplified Action Potential Models

The basic motivations for simplifying the AP mathematical models are:

- To make computer simulation of excitation wave propagation in 3D-tissue model with complex configuration feasible.
- To find a qualitative relationship between normal AP generation and propagation.

There are at least three known approaches used to simplify AP mathematical models:

a. Based on singular perturbation theory
b. Based on clamp-experiment data
c. Based on the Van der Pole relaxation generator

In some cases, the sensitivity analysis [1] allows the introduction of some simplifications to modern sophisticated mathematical models.

5.1. Simplification of AP models using perturbation theory

Physically, the approach of singular perturbation theory [2] is based on the difference in the speed of model state variables, which allows the replacement of differential equations describing the fast variables by finite (algebraic) equations.

Let us illustrate this approach using, as an example, the Nobel AP model in the following form:

$$C\frac{dV}{dt}-(\overline{g_{Na_1}}\, m^3h+g_{Na_2})(V-V_{Na})+(g_{K_1}(V)+\bar{g}_{K_2}\, n^4)(V-V_K)+g_l(V-V_l)=I_{st} \qquad (1)$$

$$\tau_m(V)\frac{dm}{dt}=m_\infty(V)-m;\quad \tau_h(V)\frac{dh}{dt}=h_\infty(V)-h;\quad \tau_n\frac{dn}{dt}=n_\infty(V)-n$$

The gate variables m and h are much faster than variable m. Indeed, $\tau_m \approx 10^{-4}$ sec, $\tau_h \approx 10^{-3}$ sec and $\tau_n \approx 10^{-1}$ sec.

Now, let us substitute $t=\bar{t}\tau_0$ (where $\tau_0=10^{-1}$ sec) in each of the Noble model equations (1).

Then we obtain:

$$\frac{C}{\tau_0}\frac{dV}{d\bar{t}}-(\overline{g_{Na_1}}m^3h+g_{Na_2})(V-V_{Na})+(g_{K_1}(V)+\overline{g_{K_2}}\,n^4)(V-V_K)+g_l(V-V_l)=I_{st}$$

$$\frac{\tau_m}{\tau_0}\frac{dm}{d\bar{t}}=m_\infty(V)-m;\quad \frac{\tau_h}{\tau_0}\frac{dh}{d\bar{t}}=h_\infty(V)-h;\quad \frac{\tau_n}{\tau_0}\frac{dn}{d\bar{t}}=n_\infty(V)-n;$$

The values $\frac{\tau_m}{\tau_0}=\varepsilon_1$; $\frac{\tau_h}{\tau_0}=\varepsilon_2$ are the small parameters and are much smaller than $\frac{\tau_n}{\tau_0}$.

When ε_1 and ε_2 are $\to 0$, the equations for variables m and h are reduced to:

$m_\infty(V)-m=0$ *and* $h_\infty(V)-h=0$. The Noble model (1) is transformed into a system of second order ODE's:

B.Ja. Kogan, *Introduction to Computational Cardiology: Mathematical Modeling and Computer Simulation*, DOI 10.1007/978-0-387-76686-7_5,

$$C\frac{dV}{dt}-(\overline{g_{Na_1}}\, m_\infty^3 h_\infty + g_{Na_2})(V-V_{Na})+(g_{K_1}(V)+\bar{g}_{K_2}\, n^4)(V-V_K)+g_l(V-V_l)=I_{st} \tag{2}$$

$$\tau_n(V)\frac{dn}{dt}=n_\infty(V)-n$$

The equations (2) are derived assuming that the time constants τ_m and τ_h are negligibly small in comparison to τ_n. Therefore, its solutions are correct for the time $t >> \tau_m, \tau_h$.

The transition from (1) to (2) (replacing m with $m_\infty(V)$ and h with $h_\infty(V)$) decreases the maximal value of the fast sodium current. The effect of this simplification on the shape of action potential is shown in Fig. 1.

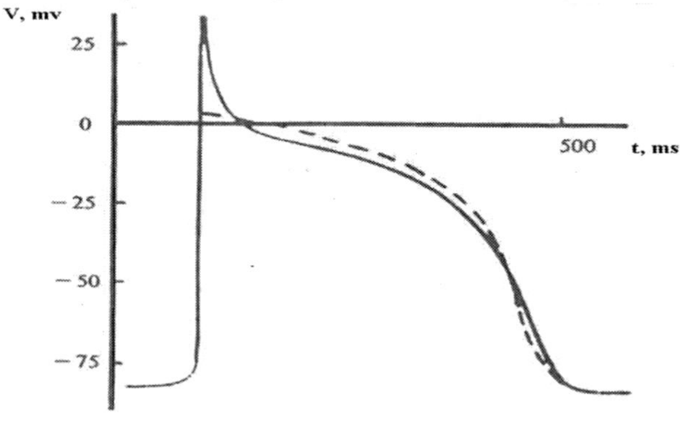

Fig. 1. Comparison of the shape of the AP calculated using the complete (solid line) and the simplified (dotted line) Noble models, taken from [3].

This effect can be partly compensated by decreasing the membrane capacity, C.

5.2. Simplification using clamp-experiment data

The single cell clamp- experiments give a family of functions:

$I_{total} = f(t, V_i)$

V_i is the fixed membrane potential ranging from V_{th} to V_{max} and I_{total} is the total current measured at the fixed membrane potential.

One of the functions from this family is shown in fig. 2.

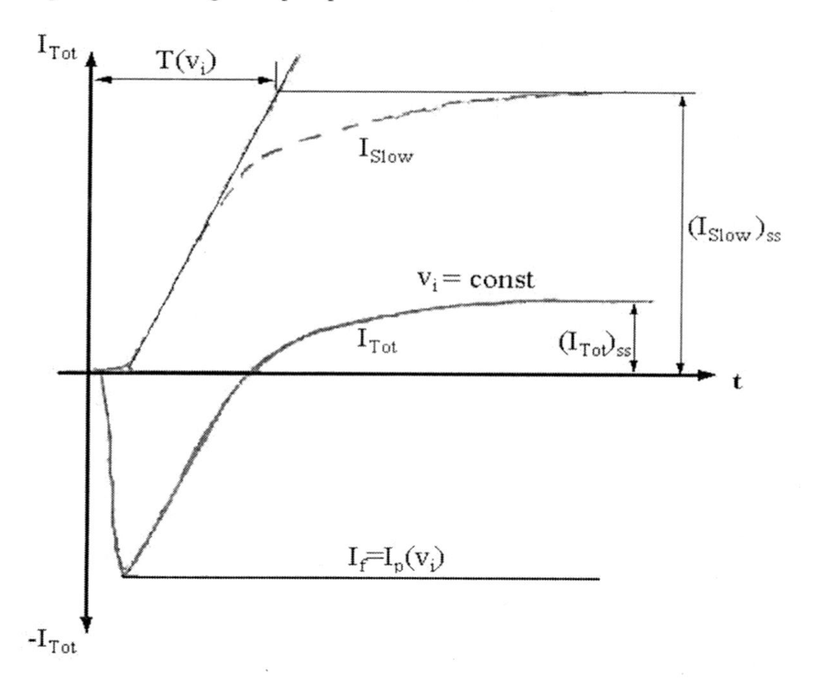

Fig. 2. Splitting of the total clamp experiment currents on fast and slow components.

According to Kirchoff's law:

$$C\frac{dV}{dt} + I_{total} = I_{st} \tag{3}$$

Let us divide I_{total} into two components: fast I_f and slow I_S. Thus, according to fig. 2, $I_f \leq 0,\ I_S \geq 0$. So $I_{total} = -I_f + I_s$ and the fast component can be expressed as:

$$I_f = \begin{cases} I_{total}(V_i) & \text{when } \dfrac{dI_{total}}{dt} \leq 0 \\ I_p(V_i) & \text{otherwise} \end{cases}$$

Equation (3) can be rewritten as: $C\frac{dV}{dt^*} = -[I_f(V) + I_S(t,V)] + I_{st}$.

Here: $t^* = t - \Delta t(V)$, $\Delta t(V)$ is the time of $(I_{total})_{\max}$

In order to find the equation for I_S, we assume that it can be represented as an exponential function of time. Therefore:

$$\tau_S(V)\frac{dI_S}{dt^*} = [I_S(V)]_{ss} - I_S$$

$$[I_S(V)]_{ss} = [I_{total}(V)]_{ss} - I_p(V)$$

Here subscript *ss* indicates the steady state value.

Taking into consideration the signs of the currents and neglecting $\Delta t(V)$, the final version of the simplified mathematical model will be:

$$C\frac{dV}{dt} = I_p(V) - I_S(t,V) + I_{st} \qquad (4)$$

$$\tau_S(V)\frac{dI_S}{dt} = [I_S(V)]_{ss} - I_S$$

All the functions in (4) are defined by tables of clamp-experiment data.

5.3. From Van Der Pol to FitzHugh-Nagumo simplified model

5.3.1. Preliminary considerations

A second-order linear R, L and C oscillator is described in Chapter 2 by the following differential equation:

$$\frac{d^2V}{dt^2} + 2\alpha\frac{dV}{dt} + \omega_0^2 V = 0 \qquad (5)$$

with the initial conditions $V(0) = V_1$ and $(dV/dt)_{t=0} = 0$.

Analysis of the roots of this characteristic equation shows that the solution is stabile when $\alpha > 0$. Note that when $\alpha = 0$ the system will be on the border of stability, and when $\alpha < 0$ the solution is unstable.

Let us consider the state variables corresponding to equation (5) and analyze its behavior on a phase-plane.

5.3.2. State variable representation

Let us make the following substitutions in (5):

$$V = x_1; \quad \frac{dV}{dt} = x_2$$

Then (5) is reduced to a system of two first order ordinary differential equations:

$$\frac{dx_2}{dt} = -2\alpha x_2 - \omega_0^2 x_1 \qquad (6)$$

$$\frac{dx_1}{dt} = x_2 \qquad (7)$$

Variables x_1 and x_2 are called the state variables. They fully determine the state of the considered dynamical system.

5.3.3. Phase-plane approach

Let us construct a phase-plane in rectangular coordinates which abscissa represents the state variable x_1 and ordinate the state variable x_2. To derive the solution of (6) and (7) on this phase-plane let us divide equation (6) by (7):

$$\frac{dx_2}{dx_1} = -\frac{2\alpha x_2 + \omega_0^2 x_1}{x_2} \tag{8}$$

The trajectory of the solution $x_2 = f(x_1)$ in the phase-plane can be found, for some cases, by the direct integration of (8) with given initial conditions.

For example, when in (8) $\alpha = 0$ we obtain:

$$\frac{dx_2}{dx_1} = -\frac{\omega_0^2 x_1}{x_2}$$

After integration from $x_2(0)=0$ to x_2 and $x_1(0)$ to x_1:

$$\int_{x_2(0)=0}^{x_2} x_2 dx_2 = -\omega_0^2 \int_{x_1(0)}^{x_1} x_1 dx_1 \quad \text{or} \quad x_2^2 = -\omega_0^2 (x_1^2 - x_1^2(0))$$

Finally:

$$\frac{x_2^2}{x_1^2(0)\omega_0^2} + \frac{x_1^2}{x_1^2(0)} = 1 \tag{9}$$

The equation (9) represents an ellipse in the phase-plane.

For $\omega_0 = 1$, we obtain acircle with radius $x_1(0)$. The typical phase-plane plots for second order linear dynamic systems are shown in Fig. 3a,b.

Name	Roots	Sketch
Stable focus or spiral	Damped complex conjugate 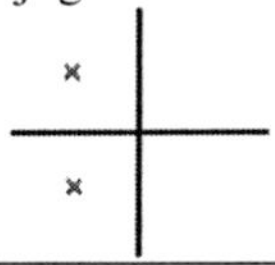	Trajectories spiral asymptotically to focus 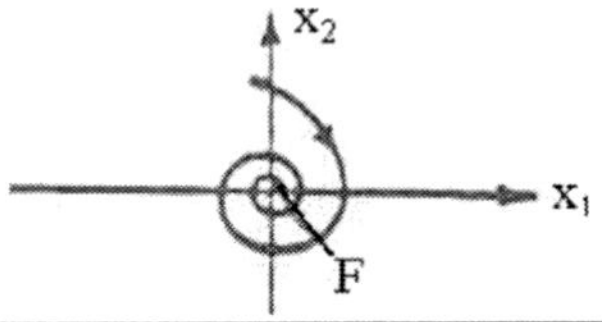
Stable node	Stable real roots 	Trajectories approach node monotonically 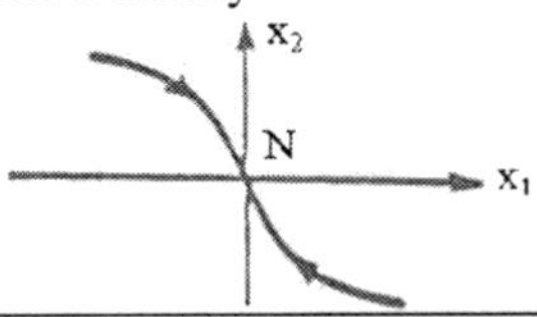
Vertex or center (Structurally unstable)	Imaginary roots 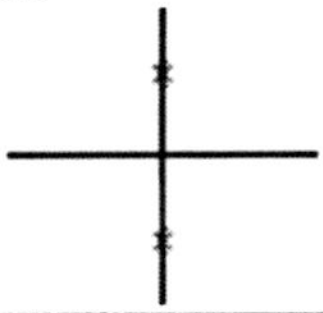	Conservative system or oscillator 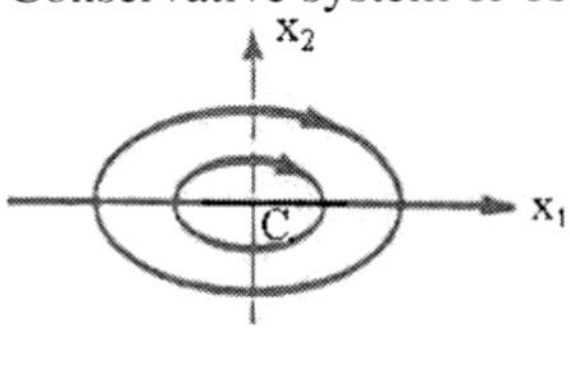

Fig. 3a. Typical phase-plane trajectories for linear second-order dynamical systems

Name	Roots	Sketch
Unstable focus	Complex conjugate with positive real part 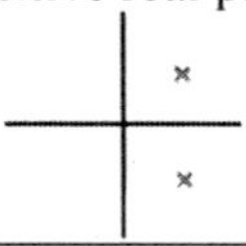	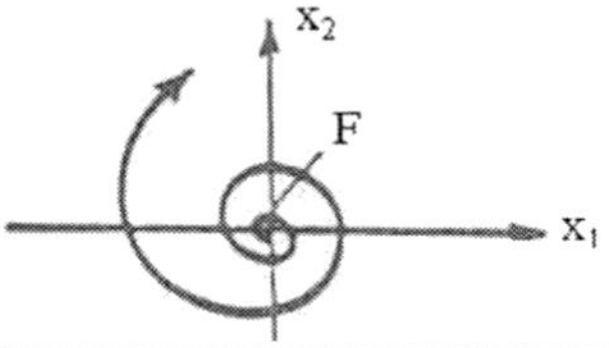
Saddle point	Unstable equilibrium point 	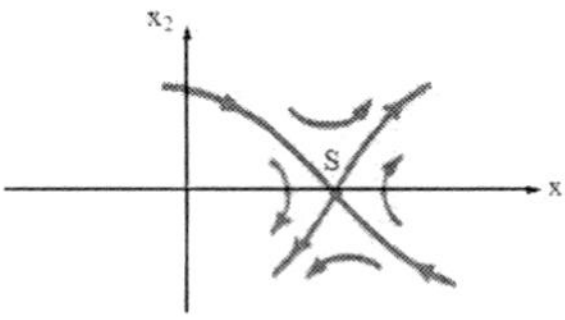
Unstable node	Unstable real roots 	Trajectories diverge monotonically from node 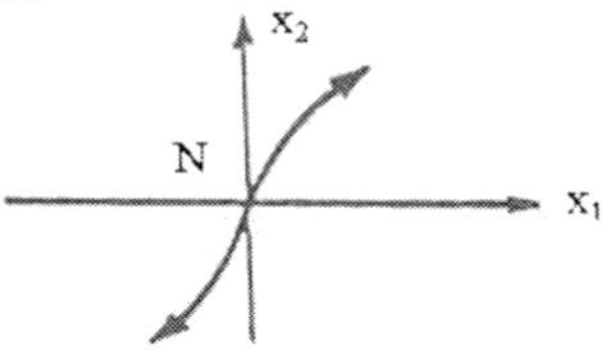

Fig. 3b. Typical phase-plane trajectories for linear second-order dynamical systems

In most cases, the direct analytical approach (to obtain the phase trajectory) cannot be applied because the variables x_1 and x_2 are not separable. Thus, the graphical and graph-analytical methods are used. Among them the method of isoclines is widely used. Isoclines are lines in phase-plane at each point of which the slope of tangent to the phase trajectory is the same.

The equation (8) $\frac{dx_2}{dx_1}$ represents a tangent to the trajectory of the solution in a phase-plane in a given point. Denoting: $\frac{dx_2}{dx_1} = \zeta = const.$, we obtain the equation for isoclines:

$$-\left(\frac{2\alpha x_2 + \omega_0^2 x_1}{x_2}\right) = \zeta \quad \text{or} \quad x_2 = -\frac{\omega_0^2}{2\alpha + \zeta} x_1 \tag{10}$$

The expression (10) represents a straight line in phase plane. The trajectories of the solutions will cross this straight line with the same inclination ζ.

Among all isoclines, the theory of oscillation distinguishes two: with $\zeta = 0$ and $\zeta = \infty$. They are called the horizontal and vertical null-isoclines and reflect the fact that $\frac{dx_2}{dt}$ or $\frac{dx_1}{dt}$ becomes zero for some points in phase-plane.

The points in the phase-plane where these derivatives become zero simultaneously are called singular points. They correspond to the equilibrium states of the considered dynamical system because the phase coordinates x_2^k and x_1^k do not change in time at these points k. The equilibrium states can be stable or unstable.

5.3.4. Relaxation Oscillations

Let us consider the Van Der Pol equation mentioned in Chapter 2 when parameter $\alpha >> \omega$:

$$\frac{d^2V}{dt^2} - 2\alpha(1-V^2)\frac{dV}{dt} + \omega_0^2 V = 0 \tag{11}$$

The solution of this equation represents periodical but not sinusoidal oscillations with period T= R_{eff} C shown in Fig. 4 according to [4].

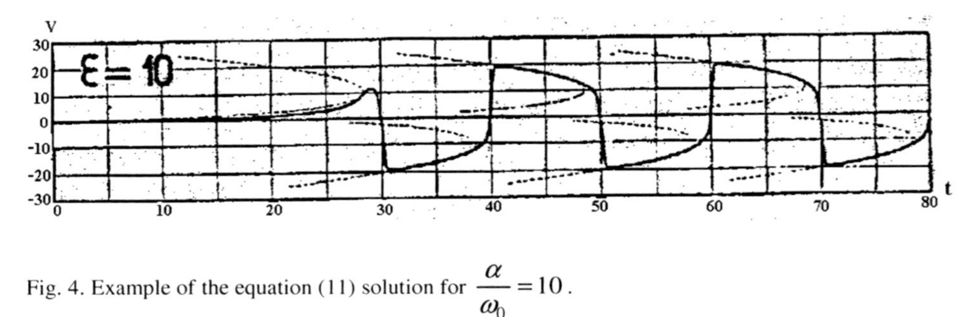

Fig. 4. Example of the equation (11) solution for $\frac{\alpha}{\omega_0} = 10$.

In this case, the solution represents a "relaxation oscillation," which is significantly different from a sinusoidal curve. Sudden jumps are seen to occur periodically.

5.3.5. Phase-Plane approach for analysis of V-P equation.

Let us introduce a new time $\tau = 2t\alpha$ in (11) and designate as a small parameter $\phi = \frac{\omega_0^2}{4\alpha^2}$. As a result, equation (11) becomes:

$$\frac{d^2V}{d\tau^2} + (V^2-1)\frac{dV}{d\tau} + \phi V = 0 \tag{12}$$

In order to study (12) on a phase-plane this equation must be reduced to a system of two first order ordinary differential equations. According to the Lienard transformation, we introduce state variables $W = -\frac{dV}{d\tau} + V - \frac{V^3}{3}$ and V.

Thus,

$\frac{dW}{d\tau} = -\frac{d^2V}{d\tau^2} + \frac{dV}{d\tau}(1-V^2)$. The right side of this equation, according to (12), is equal to ϕV. Therefore,

$$\frac{dW}{d\tau} = \phi V \tag{13}$$

The ODE for the second state variable V is obtained from expression for W, solving it in respect to $\frac{dV}{d\tau}$.

$$\frac{dV}{d\tau} = V - \frac{V^3}{3} - W \tag{14}$$

We will study the solution of equations (13) and (14) on the phase plane W, V using the method of isoclines. In particular, let us consider the null-isoclines.

The equation for horizontal and vertical null - isoclines follows from (13) and (14) (when $\frac{dW}{d\tau}$ and $\frac{dV}{d\tau}$ are set to zero). Indeed,

For horizontal null-isoclines:

$$\phi V = 0 \tag{15}$$

For vertical null-isoclines:

$$W = V - \frac{V^3}{3} \tag{16}$$

The equations (15) and (16) for null-isoclines on a phase-plane W, V describe the ordinate axis and a cubic parabola respectively (see Fig.5).

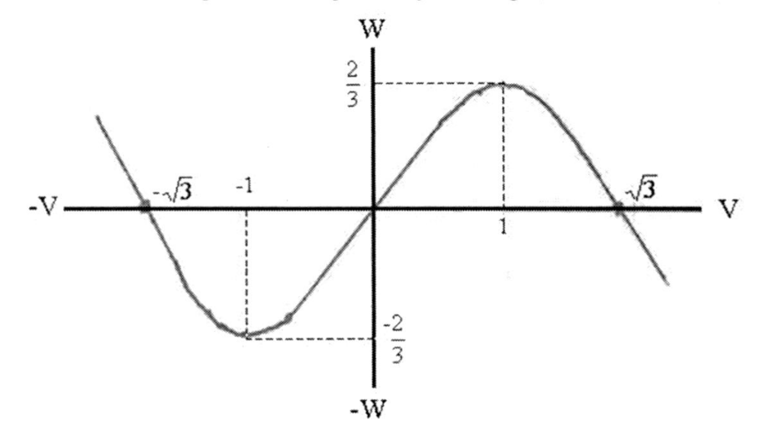

Fig. 5. Null-isoclines for Van Der Pol equation

The origin of the coordinate systems is the singular point where $\frac{dW}{d\tau} = \frac{dV}{d\tau} = 0$. Let us deteremine if this singular point is stable or unstable.

The stability of the steady state point is determined by stability of the system that is in close proximity to this point or in other words, by the stability in response to small perturbations.

Let us designate the small deviations from the steady state point ΔW and ΔV. Introducing these small deviations in (13) and (14) and neglecting $\frac{(\Delta V)^3}{3}$ in comparison to V, we obtain:

$$\frac{d\Delta W}{d\tau} = \phi \Delta V$$

$$\frac{d\Delta V}{d\tau} = \Delta V - \Delta W$$

This is a system of linear differential equations. The stability of its solution depends of the location of the roots of a corresponding characteristic equation in the complex plane. The characteristic equation is:

$$D^2 - D + \phi = 0$$

The roots of this quadratic equation are:

$$D_{1,2} = +\frac{1}{2} \pm \sqrt{\frac{1}{4} - \phi} \quad \text{where } \phi \approx 0.01 - 0.08$$

Both roots are real, positive and different and the solution of our original system for small perturbation around singular point will be unstable. Such a singular point in phase-plane is called the unstable center.

Determination of the direction of the tangent to phase trajectories on horizontal and vertical Null-Isoclines

The expression for a tangent ς to the trajectory of the solution to the VP equation on the phase plane is:

$$\frac{\frac{dW}{d\tau}}{\frac{dV}{d\tau}} = \frac{\phi V}{V - \frac{V^3}{3} - W} = \varsigma = tg\beta \qquad (17)$$

β is the angle between the positive V axis and tangent to a phase trajectory.

For horizontal null-isoclines $\frac{dW}{d\tau} = \phi V = 0$ so $\varsigma = tg\beta = 0$ and $\beta = 0$ or 180°.

Let us determine the direction of the tangent for positive and negative parts of the axis *W*. For this purpose assume that ϕV is a very small but finite value equal to ε_1. According to (17) and when

W>0, then $tg\beta = -\frac{\varepsilon_1}{W}$ and $\beta \cong 180^0$.

W<0, then $tg\beta = \frac{\varepsilon_1}{W}$ and $\beta \cong 0^0$.

The corresponding marking of the axis *W* is shown in Fig. 6.

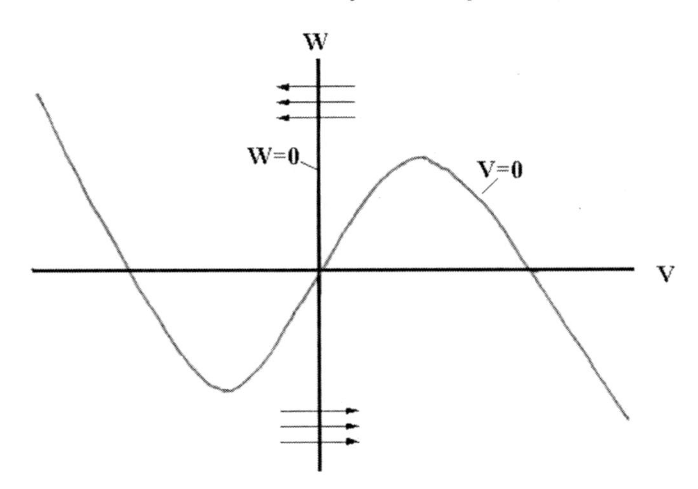

Fig. 6. Horizontal null isoclines for VP equation

For vertical null-isoclines $\frac{dV}{d\tau}=0$, according to (17), it follows that $tg\beta=\infty$ and $\beta=\frac{\pi}{2}$ or $3\frac{\pi}{2}$. In order to find the distribution of angle β along the vertical null-isoclines (in our case, a cubic parabola), let us consider $\frac{dV}{d\tau}$ as a very small positive but finite value equal to ε_2.

Then, it follows from (17) that $tg\beta=\frac{\phi V}{\varepsilon_2}$ and the sign of $tg\beta$ will coincide with the sign of *V*. So, for positive *V*, the angle β will be equal to $\frac{\pi}{2}$ and for negative *V* to $3\frac{\pi}{2}$. The corresponding marking of both null-isoclines is shown in Fig. 7

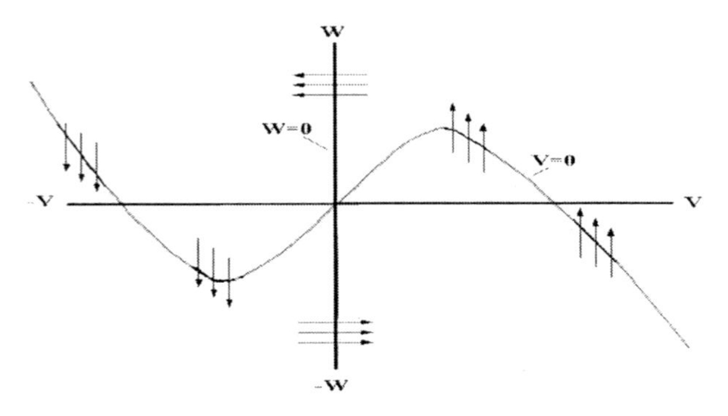

Fig. 7. The trajectories directions of solution to Van Der Pol equation on phase-plane null-isoclines

The limit cycle in phase-plane and corresponding pulse (AP) train, generated in time, are presented in Figures 8 and 9 respectively.

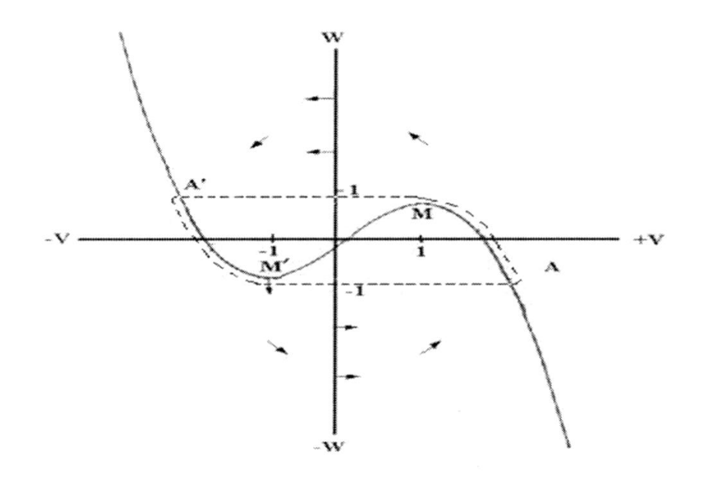

Figure 8. Trajectory of the solution on the phase plane plot.

The point in this phase-plane, which represents the equilibrium state of the system, is the origin of the coordinate where both null-isoclines cross. It was proved before that this equilibrium state is unstable. Van Der Pol used the model (13), (14) for the simulation of the heart pacemaker system.

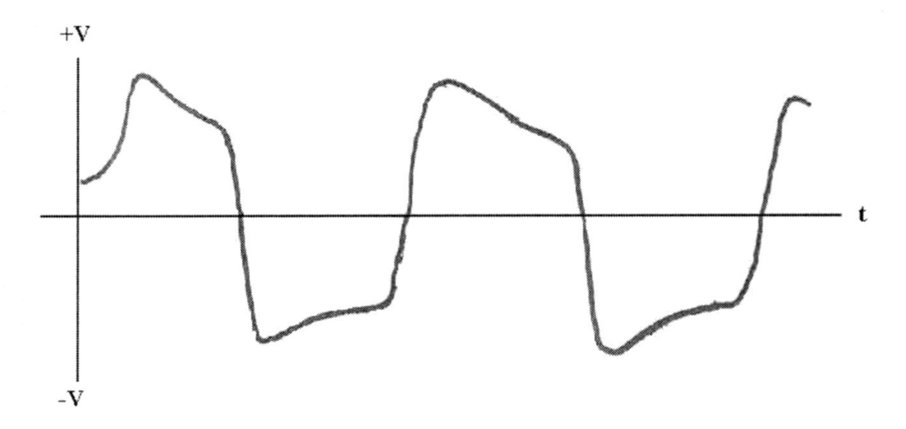

Fig. 9. The solution of the VP equation in time represents the relaxation oscillations, one of the non-sinusoidal types of oscillation with period depending of small parameter ϕ value.

5.3.6. FitzHugh modification of VP equations

FitzHugh proposed to modify the BVP (B. Van Der Pol) equations to simulate nerve AP. He changed the vertical isocline, the cubic parabola, by shifting it along the vertical coordinate W by adding step-function current I [5]. The line of horizontal isoclines is changed from the vertical axis W to a sloped straight line. All these modifications are introduced in equation (18) and shown in figs. 10, 11, 12, and 13.

$$\dot{V} = V - \frac{V^3}{3} - W + I \tag{18}$$

$$\dot{W} = \varepsilon\ (V + a - bW)$$

Here: $\varepsilon = \phi = 0.08$, a=0.7, b=0.8

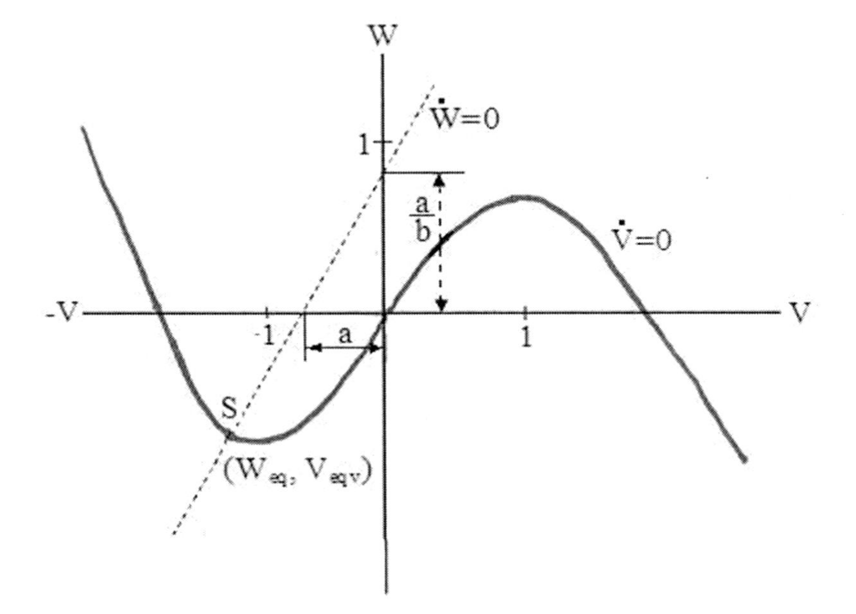

Figure 10. Null isoclines for FitzHugh modifications of BVP relaxation generator

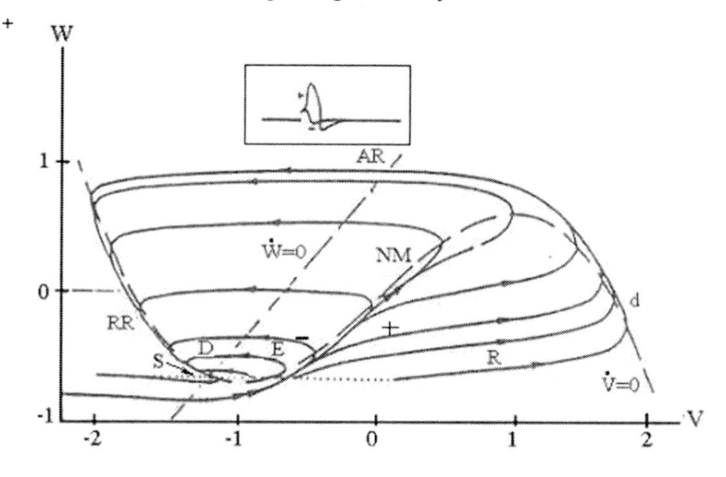

Fig. 11. Phase plane of cubic BVP modified equations (18), for instantaneous current I shocks of various amplitudes. Broken curves are null-isoclines. Shock displaces a state point from a resting point to a point on the dotted line. Double arrowheads denote the threshold separatrix. In this and subsequent figures, + and – indicate a pair of trajectories, one for a stimulus slightly above (+) and one slightly below (-) threshold. a = 0.7, b = 0.8, ε = 0.08.

Encircled letters denote the physiological states as follows: A = Active, AR = absolutely refractory, D = depressed, E = enhanced, NM = no man's land, R = regenerative, RR = relatively refractory. Inset: curves of V versus t, showing an action potential (+) and an active sub-threshold response (-).

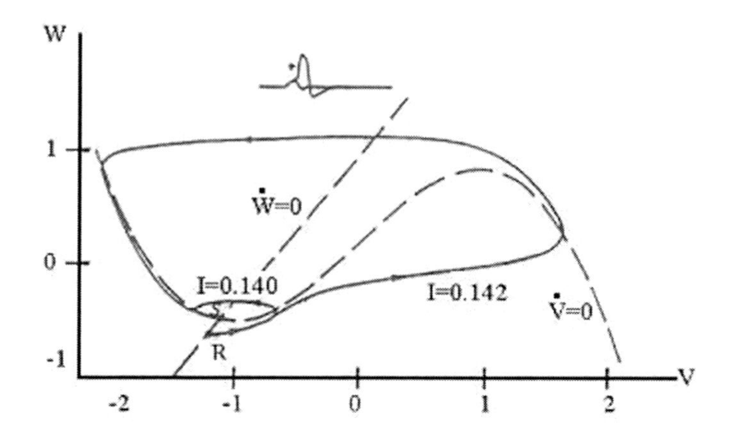

Fig. 12. The phase plane for modified BVP equations showing stimulation by step currents above (I = 0.142) and below (I = 0.140) rheobase.

The curve of vertical isoclines ($\dot{V}$ = 0) is raised by positive values of I. R = resting point, S= singular point.

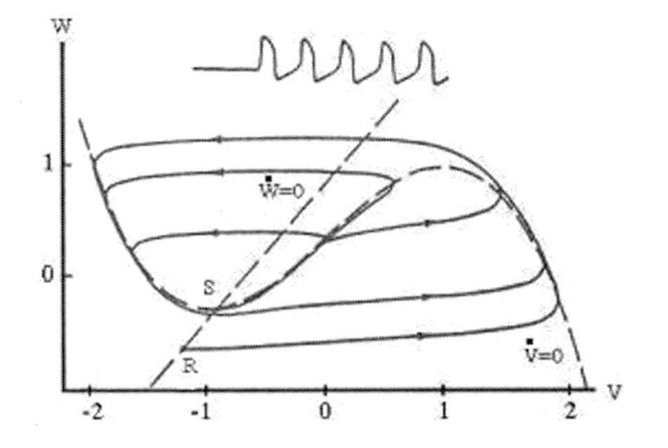

Fig. 13. BVP phase plane for step current I= 0.4, with an unstable singular point S and a plus stable limit cycle. Inset above: endless train of action potentials. R = resting point.

5.3.7. Nagumo analog model based on nonlinear properties of tunnel diode

Nagumo et. al. [6] proposed to use an electrical circuit (see Fig.14) containing a tunnel diode as an analogy model for nerve AP generation.

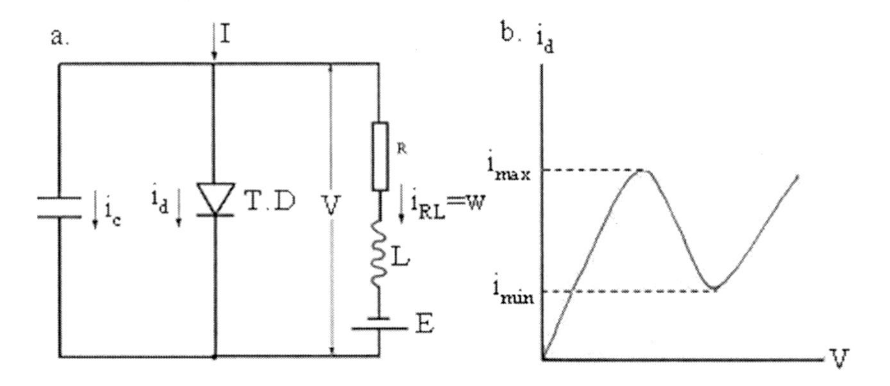

Fig. 14. (a) Equivalent circuit diagram, (b) voltage-current characteristics of a tunne diode

Kirchoff's law requires that the sum of internal currents is equal to the external current. Indeed,

$$C\dot{V} + F(V) + W = I \tag{19}$$

$$L\dot{W} + RW - E = V$$

Here $F(V)$ is current, i_d, through a tunnel diode (see Fig. 14b).

This function may be made close to the $F(V) = V - \frac{V^3}{3}$ in the Van Der Pol equation. Solving each equation of (19) with respect to its first derivative, we get: $\dot{V} = \frac{1}{C}(-F(V) - W + I)$ and $\dot{W} = \frac{1}{L}(V + E - RW)$

After reducing the above equations to dimensionless form, we get the FitzHugh equations (18) for a nerve cell. Indeed:

$$\frac{d\overline{V}}{d\overline{t}} = +\overline{F}(V) - \overline{W} + \overline{I}$$

$$\frac{d\overline{W}}{d\overline{t}} = \varepsilon(\overline{V} + \overline{E} - \overline{W})$$

Here: $\overline{V} = \frac{V}{V_m}$; $\overline{W} = \frac{W}{I_m}$; $\overline{I} = \frac{I}{I_m}$; $\overline{E} = \frac{E}{E_m}$; $\overline{t} = \frac{t}{CR_{eqv}}$; $\varepsilon = \frac{CR_{eqv}}{L/R}$; $I_m = \frac{V_m}{R}$

5.3.8. Simplification of FitzHugh-Nagumo equations for heart muscle cell

In order to investigate the FitzHugh-Nagumo equations analytically, the research group of the Biophysics Institute in Pushchino (Russia) proposed [7] to replace the function *F(V)*, a cubic parabola, by its piece-wise linear approximation and constant small parameter $\varepsilon(V)$ with step wise approximation of *V* as it is shown in Fig. 15.

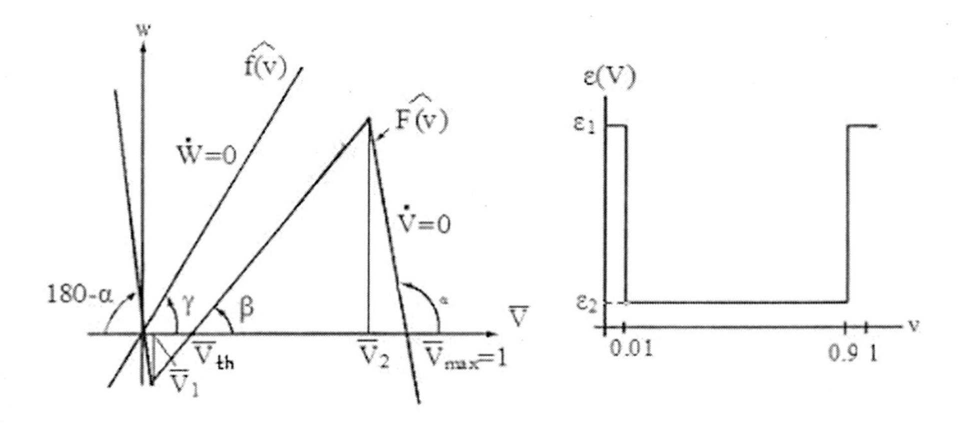

Fig.15. Null isoclines and ε(V) plot for the FitzHugh-Nagumo model with a linear approximation.

All parameters of this model were chosen to roughly reproduce the major characteristics of cardiac normal AP and are presented below as in [7].

$\tan\beta = G_f$; $\tan\gamma = G_s$; $\tan\alpha = G_r$

$G_f = 1$; $G_s = 1$; $G_r = 30$; $V_{th} = 0.16$;

$$\varepsilon(V) = \begin{cases} 0.5 & \text{when} \quad 0 < V < 0.01 \\ 0.01 & \text{when} \quad 0.01 \le V \le 0.95 \\ 0.5 & \text{when} \quad V > 0.95 \end{cases}$$

The simplified model equations have the form:

$$\frac{d\overline{V}}{d\bar{t}} = \hat{F}(V) - \overline{W} + I_{st}$$

$$\frac{d\overline{W}}{d\bar{t}} = \hat{\varepsilon}(V)[\hat{f}(V) - \overline{W}] \tag{21}$$

The parameters of the model must be chosen to provide the same relaxation coefficient $\rho = \frac{D}{\Delta t} = \frac{D\dot{V}_m}{V_m}$ as in a real cardiac cell A.P.(see Fig. 16).

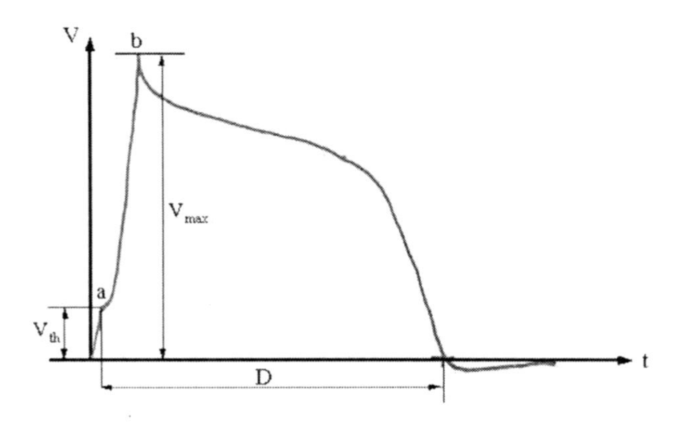

Fig. 16. To the determination of relaxation coefficient ρ

5.4. Comparative analysis of the second order simplified AP models

5.4.1. General comments

The comparative analysis of the three most widely used simplified AP models: FitzHugh-Nagumo, Van Capelle-Durrer and Karma was first presented in [8]. In this section we demonstrate a common approach to synthesis of simplified second order AP models.

From here on, the total ionic current is divided into two components: inward current I_{inw}, which flows from outside to inside the cell, and outward current I_{outw}, which flows from inside to outside the cell. The simplification of the full mathematical model of AP generation is based on two assumptions:

1. The gate variables m and h, which determine the conductivity of the membrane for the sodium current in the full model, are changing so rapidly that it is possible to assume that they reach their steady state values $m_\infty(V)$ and $h_\infty(V)$ instantaneously. For the Noble model, it was proven in [3], using singular perturbation theory, that this assumption is correct. This assumption causes a decrease of the maximal rate of depolarization and can be easily compensated by decreasing the membrane capacity or by increasing the maximum value of the inward current.
2. The total outward current being a function of membrane potential, V, and time, t, can be represented as a product of two functions: a function of one variable V, representing the dependence of steady state generalized outward current on membrane potential V, and a function of two variables V and t, representing the generalized gate variable for the generalized outward current: $I_{outw} = Y^k(V, t)\, I_{outw}(V)$.

Here, Y is the dimensionless generalized gate variable for outward current ($0 \leq Y \leq 1$), and k is a constant; usually k is chosen greater than unity to minimize the effect of outward current on the depolarization processes. For example, in the Noble equations $k = 4$.

The second assumption excludes the time independent components of outward current inherent in physiological models (see [9] and [10]). These components are responsible for the first part of the repolarization phase of AP and, particularly, for the overshoot. Therefore, this assumption introduces additional errors in the reproduction of the AP shape.

The generalized gate variable $Y(V,t)$ is obtained as a solution of the equation:

$$\tau_Y \frac{dY}{dt} = \mathrm{Y}_\infty(V) - \mathrm{Y} \tag{22}$$

The Noble equations show that only one gate variable n does not reach the steady state value after the completion of the repolarization processes. The transient of that gate variable during diastole determines the APD restitution curve, a very important determinant of repeatable wave propagation. This transient cannot be described properly by (22) with τ_Y being a constant. τ_Y must be a function of V and $[dY/dt]$:

$$\tau_Y = \begin{cases} \tau_1 & \text{if } \dfrac{dY}{dt} > 0 \\ \tau_1 K_1 & \text{if } \dfrac{dY}{dt} < 0, \text{ and } Y > Y_{\lim} \\ \tau_1 K_2 & \text{otherwise} \end{cases}$$

The value of τ_1 determines the AP duration and, under normal conditions, is 100 times greater than the time constant of the fast variable V. The coefficient K_2 can be found directly from the given APD restitution curve (see [11]), and K_1 and Y_{lim} are established in the process of the final adjustment of this restitution curve. The function $Y_\infty(V)$ has to be equal to zero when $V=V_{rest}$ and equal to 1 when $V=V_{max}$. Inside this range of $V \subset [V_{rest}, V_{max}]$, any rough stepwise or piece-wise linear approximations can be used (see, for example, fig.17). The effect of n on $[dV/dt]_{max}$ of the next AP is negligible.

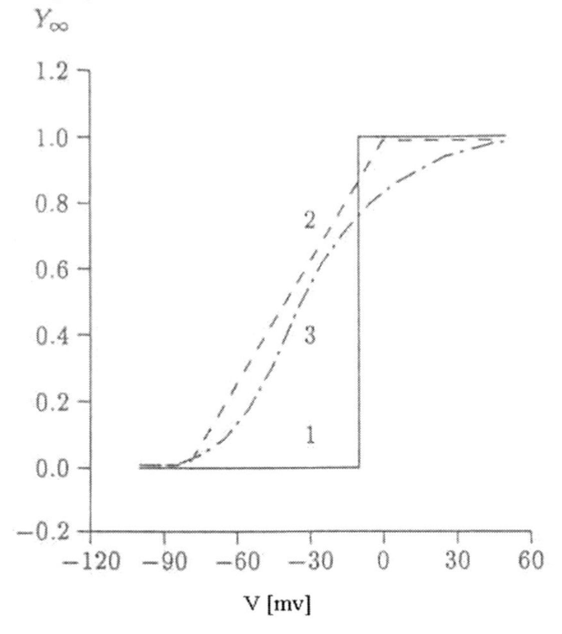

Figure 17. The $Y_\infty(V)$ dependences used in the: 1) Karma model, 2) Van Capelle-Durrer model, and 3) Noble model.

Therefore, the general form of the simplified model based on the Noble formulation [11] and the above mentioned assumptions is:

$$C\frac{dV}{dt} = -I_{inw}(V) - \left(\frac{Y}{Y_B}\right)^k I_{outw}(V) \tag{23}$$

$$\tau_Y(V)\frac{dY}{dt} = \mathrm{Y}_\infty(V) - \mathrm{Y} \tag{24}$$

with the initial conditions: $V(0) = V_{rest}$, $Y(0) = 0$. Here Y is a generalized gate variable. The introduction of the ratio $[Y/Y_B]^k$ instead of Y^k following Karma [12] permits the adjustment of parameters Y_B and k to control the action potential duration and the shape of the AP during the last part of repolarization phase.

The behavior of the models described by general equations (23) and (24) can be analyzed using $I_{inw}(V)$ and $I_{outw}(V)$ plots and some general properties of the AP. Indeed, the slope of the AP at any given point of time is completely determined by the difference between $I_{inw}(V)$ and $[Y/Y_B]^k\ I_{outw}(V)$. During the depolarization phase, $[Y/Y_B]^k \approx 0$, and the outward current does not affect the depolarization processes. If $[Y/Y_B]^k$ continues to be zero after $I_{inw}(V)=0$ or $I_{inw}(V)=[Y/Y_B]^k\ I_{outw}(V)$, we obtain a plateau without overshoot. To obtain fast repolarization, the following has to be true: $[Y/Y_B]^k I_{outw}(V) > I_{inw}(V)$.

Another form of the generalized second order simplified model can be obtained if I_{outw} in (23) is represented as the sum of two components $I_{outw} = -I_{inw} + \Delta I_{outw}$

$$C\frac{dV}{dt} = -I_{inw}(V)\left[1-\left(\frac{Y}{Y_B}\right)^K\right]-\left(\frac{dV}{dt}\right)^k \Delta I_{outw} \tag{25}$$

$$\tau_Y(V)\frac{dY}{dt} = \mathrm{Y}_\infty(V) - \mathrm{Y} \tag{26}$$

The current ΔI_{outw} can be defined (as a piecewise-linear function) to reproduce the desired shape of the repolarization phase of the AP. I_{inw} can be obtained from experimental data.

Using these general considerations let us analyze the existing simplified models.

5.4.2. FitzHugh-Nagumo model

The FitzHugh-Nagumo (FH-N) model is the simplest of the existing simplified models. The mathematical model equations, reduced to dimensionless form [5, 6] are:

$$\frac{dV}{dt} = +F(V) - I + I_{stim}$$

$$\frac{dI}{dt} = \varepsilon(V)\left[f(V) - I\right] \tag{27}$$

Here

- V - fast variable (membrane potential displacement between the interior and exterior domains of the cell)
- I - slow variable (generalized outward current)
- I_{stim}- stimulus current,
- F(V) - current-voltage characteristic of the fast inward current
- f(V) - current-voltage characteristic of the slow outward current
- ε(V) - small parameter (inversely proportional to the time constant of the slow outward current).

The piece-wise linear approximation of the functions $F(V)$, $f(V)$, and $\varepsilon(V)=[1/(\tau Y(V))]$ are shown in Fig. 18.

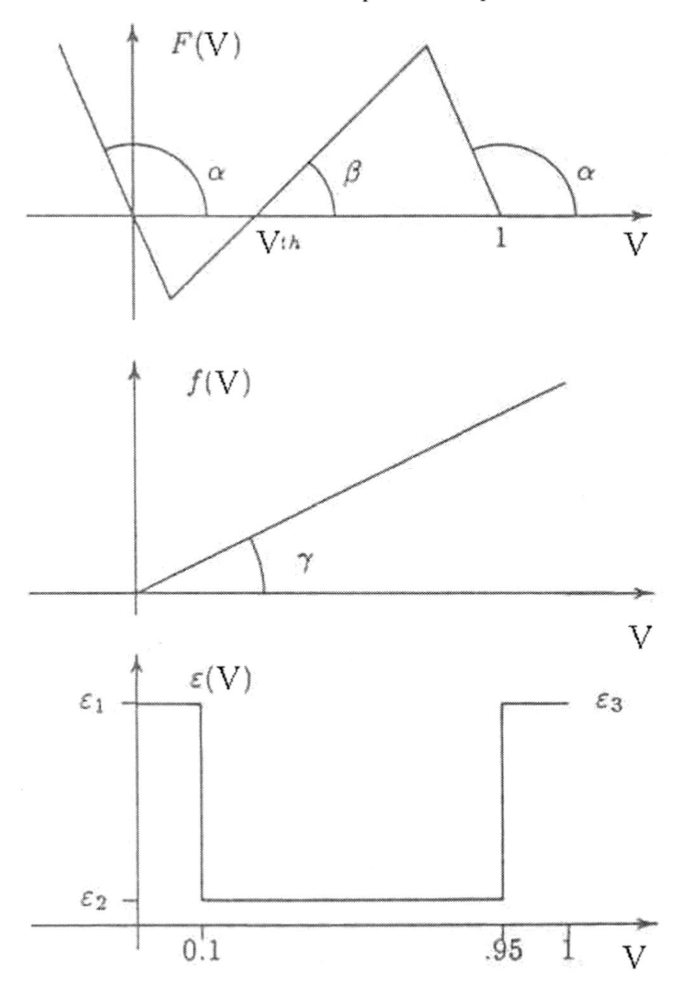

Fig.18. The piece-wise-linear approximation of functions F (V), f (V), and ε(V). The tan $\alpha = G_r$; tan $\beta = G_f$; tan $\gamma = G_S$; V_{th} is the threshold potential.

To solve equations (27), it is necessary to choose the appropriate initial and boundary conditions.

Let us reduce equations (27) to the general form given by (23) and (24). After introducing the substitutions $I = Yf_1(V)$ and $f(V) = Y_\infty(V)f_1(V)$ in (27) we obtain:

$$\frac{dV}{dt} = +F(V) - Yf_1(V) + I_{stim}$$

$$\frac{dY}{dt} = \varepsilon(V)\left[Y_\infty(V) - Y\right] \tag{28}$$

Here $Y_\infty(V)$ is a step function of V; $F(V)$ and $Yf_1(V)$ can be considered the generalized inward and outward currents. In Fig. 19, these currents are shown as functions of V. The graph shows that after the end of the depolarization phase, the action potential will have an abnormally prolonged plateau phase and a short fast repolarization phase.

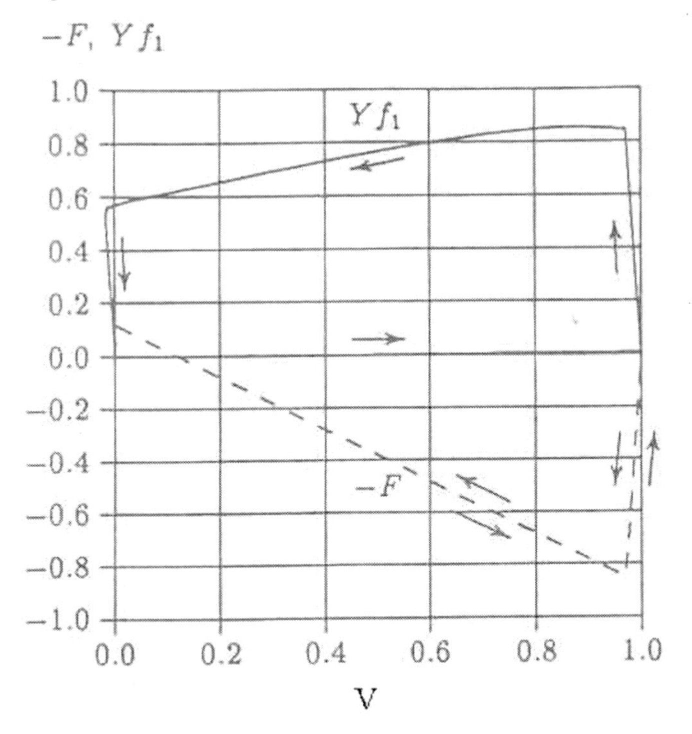

Fig. 19. Functions *F(V)* and *Y* $f_l(V)$ for FH-N model. The arrows pointing from left to right designate the changes of inward (–*F(V)* and outward (*Y* $f_l(V)$) currents in the depolarization phase of the AP, and arrows pointing in the opposite direction show the changes of the same currents in the repolarization phase.

The widely used set of parameters for this model (see [13-18]), hereafter referred to as the standard set, is: G_s= 1 ; G_f = 1 ; G_r = 30 ; V_{th}= 0.16 and

$$\varepsilon(V)=\begin{cases}\varepsilon_1 & if\ \ 0.00<V<0.01\\ \varepsilon_2 & if\ \ 0.01<V<0.95\\ \varepsilon_1 & if\ \ V>0.95\end{cases}\quad \text{, where } \ \varepsilon_1=0.5 \text{ and } \varepsilon_2=0.01$$

The relationships between the action potential duration (APD), refractory period (R), and the model parameters G_s , G_f, and ε are presented in [15] for the model without diffusion (point model). These dependencies qualitatively reflect the essential properties of heart muscle cells, but do not correctly express the restitution properties and the shape of the AP.

The transients of *V* and *Y* in a cardiac cycle with a small DI are shown in fig. 20 for the standard parameters of the F-N simplified model. Note that the slow variable *Y* reaches zero almost immediately after the AP returns to the resting potential. Thus, a stimulus applied after a short DI produces a subsequent AP of approximately the same duration as the previous one, because the gate variable Y almost decreases instantaneously when $[dY/dt] < 0$ (see Fig. 20). The original APD restitution curve in the FH-N model is very steep (see Fig. 21 curve 1) and differs quite a bit from the experimental curve (Fig. 21 curve 3). Therefore a method was developed [19] to modify the function $\varepsilon(V)$ to fit the experimental APD restitution curve.

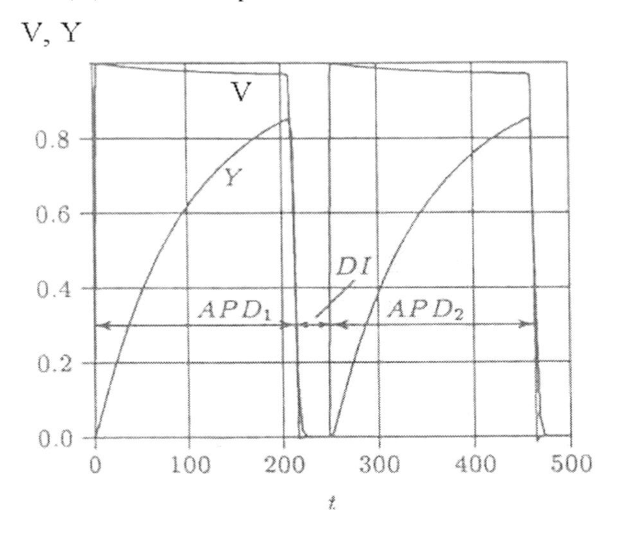

Fig. 20. The $V(t)$ and $Y(t)$ for short diastolic interval. DI/APD1=0.234 $APD_2/APD_1=1$ (in model [15])

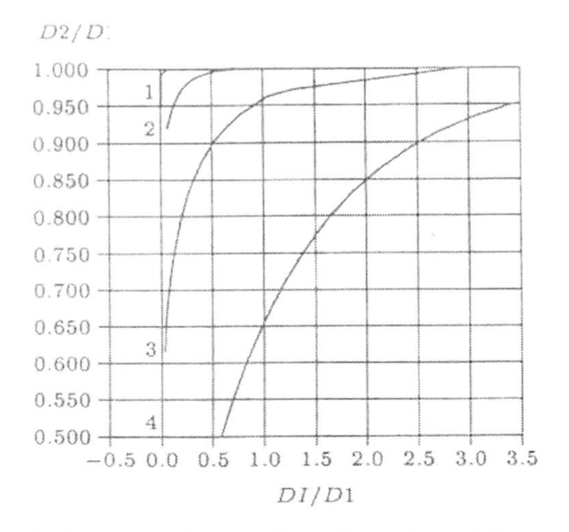

Fig.21. The APD restitution curves in normalized form obtained for the original models: 1) FitzHugh-Nagumo, 2) Van Capelle-Durrer, 3) Lou-Rudy, and 4) Karma.

According to this method (see [19] for details) the previous expression for $\varepsilon(V)$ can now be written as:

$$\varepsilon(V) = \begin{cases} \varepsilon_1 & \text{if } V < 0.01 \text{ and } \dfrac{dY}{dt} > 0 \\ \varepsilon_2 & \text{if } V \geq 0.01 \text{ and } \dfrac{dY}{dt} > 0 \\ \varepsilon_3 & \text{if } Y > Y_{\min} \text{ and } \dfrac{dY}{dt} \leq 0 \\ \varepsilon_4 = k\varepsilon_2 & \text{otherwise} \end{cases}$$

5.4.3. Van Capelle-Durrer model

The Van Capelle and Durrer model [20] can be derived from general equations (25) - (26) as follows: we rename I_{inw} as $-f$ and ΔI_{outw} as i_1, and then replace the term $1-[Y/Y_B]^k$ in equation (25) with $(1-Y)$, replace the term $[Y/Y_B]^k$ with 1, and set $\tau Y(V) = T$ = constant in equation (26). As a result, we get:

$$C\frac{dV}{dt} = (1-Y)f(V) - i_1(V) \tag{29}$$
$$T\frac{dY}{dt} = Y_\infty(V) - Y$$

The functions $(1-Y)f(V)$ and $i_1(V)$ are shown in fig. 22 in the form proposed by VCD [20]. The figure clearly shows that the maximal rate of $[dV/dt]$ in the depolarization phase can be determined by:

$$\left(\frac{dV}{dt}\right)_{\max} = \frac{(1-Y)f(V_a) - i_1(V_a)}{C} \quad \text{or}$$

$$\left(\frac{dV}{dt}\right)_{\max} = \frac{f(V_a)}{C} - \frac{Yf(V_a) + i_1(V_a)}{C} \tag{30}$$

V_a is the value of membrane potential at which $[dV/dt]$ reaches its maximum value. The first term on the right side of equation (30) represents the ideal value of $([dV/dt])_{max}$, and the second term is the error introduced by the VCD model. The error grows with the increasing values of Y and $i_1(V_a)$. This occurs in the case of repetitive stimulation (an increase of Y) and when a decrease of APD is achieved by an increase of i_1.

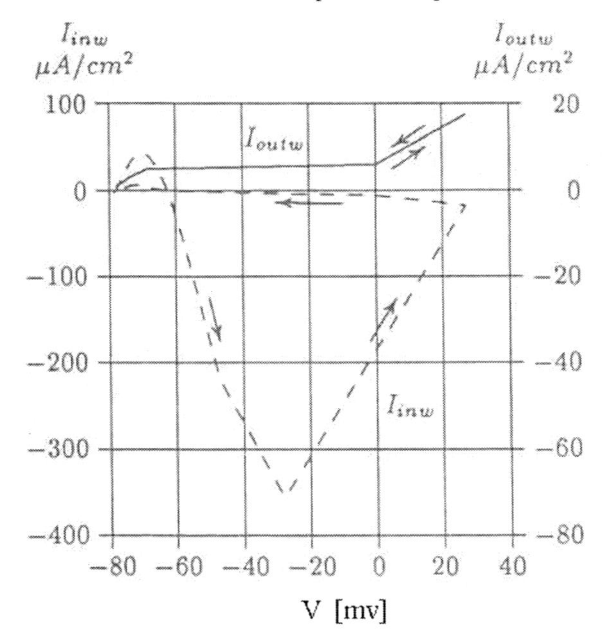

Fig. 22. I_{inw}=$(1\text{-}Y)f(V)$ and I_{outw}=$i_1(V)$ currents for the VCD model. The arrows pointing from left to right designate the changes of inward and outward currents in the depolarization phase of AP and arrows pointing in the opposite direction show the changes of the same currents in the repolarization phase.

The replacement of $\tau_Y(V)$ with T = const leads to an error in the reproduction of APD restitution properties (see Fig. 21 curve 2). This error and the error in $([dV/dt])_{max}$, inherent in the VCD model, affect the conduction velocity and, therefore, restrict the domain of the applicability of this model.

The transient of the AP and gate variable *Y* for the VCD model are shown in fig. 23.

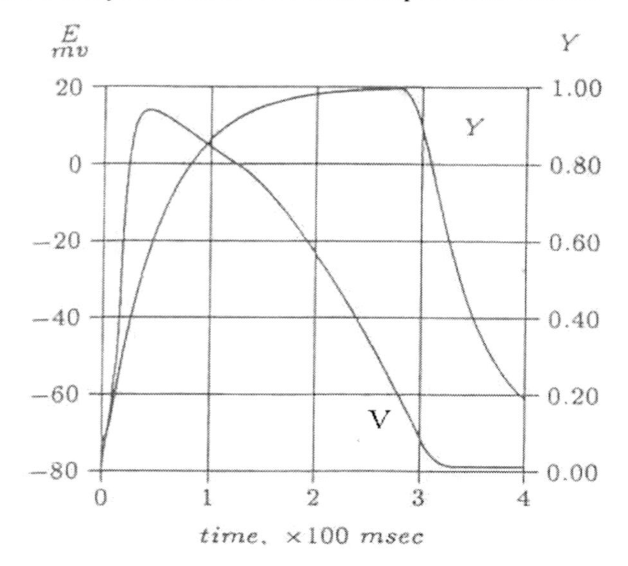

Fig. 23. The AP and gate variable Y as functions of time in the VCD model.

5.4.4. Karma model

This model [12] is a simplification of the Noble model [11]. It has three essential properties of the Noble model that are missing in the original FitzHugh-Nagumo model:

1. Insensitivity of $([dV/dt])_{max}$ to changes in the slow gate variable n in the case of repeatable excitation
2. The fast repolarization period is much longer than that of the fast depolarization period
3. There is returned alternation in APD in the case of repeatable excitation.

The two-variable model equations in the dimensionless form proposed by Karma [12] are:

$$\varepsilon\frac{\partial V}{\partial t} = \varepsilon^2\nabla^2 V - V + \left[A - \left(\frac{n}{n_B}\right)^M\right]\times[1-\tanh(V-3)]\frac{V^2}{2}$$

$$\frac{\partial n}{\partial t} = \theta(V-1) - n \qquad (31)$$

where

- $\theta(x)$ is the standard Heaviside step-function
- ε is a small parameter characterizing the abruptness of excitation
- A is a constant
- ∇^2 is a Laplacian of appropriate dimension of the tissue
- n is a slow gate variable equivalent to Y in (25).

Typical values of the parameters are $A = 1.5415$, $M = 30$, $n_B = 0.507$, $\varepsilon = 0.009$.

Let us reduce the equations (31) to the generalized form. For this purpose, consider the isolated cell $\nabla^2 V = 0$. After defining $f(V) = [1 - \tanh(V-3)]\times[V^2/2]$ and $n_\infty(V) = \theta(V-1)$, and substituting $\bar{t} = [t/\varepsilon]$; $T = [1/\varepsilon]$ into (31) we finally get:

$$\frac{dV}{d\bar{t}} = [Af(V) - V] - \left(\frac{n}{n_B}\right)^M F(V)$$

$$T\frac{dn}{d\bar{t}} = n_\infty(V) - n \tag{32}$$

In these equations ($A\,f(V) - V$) corresponds to the inward current and $[n/n_B]^M F(V)$ to the outward current. So,

$$\frac{dV}{dt} = I_{inw}(V) - I_{outw}(V,t)$$

$$T\frac{dn}{dt} = n_\infty(V) - n \tag{33}$$

In fig. 24, $I_{inw}(V) = -(Af(V)-V)$ and $I_{outw}(V, t) = ([n/(n_B)])^M f(V)$ are shown in the coarse of AP generation (fig. 25). During the depolarization phase, I_{inw} traces the bottom curve in the direction of the arrows from left to right. At the same time, $I_{outw}=0$ due to $([n/(n_B)])^M << 1$ (the arrows along x-axis). Therefore, in the case of repeatable excitation with small but finite diastolic intervals, the outward current does not affect the depolarization processes, and especially $([dV/dt])_{max}$. A growth of $[Af(V)]_{max}$ will cause a rise of $([dV/dt])_{max}$. After I_{inw} becomes equal to 0, and V reaches V_{max}, the repolarization phase begins. Due to the small excess of outward current, the membrane potential begins to decrease at a slow rate, producing in the time domain something resembling a plateau. The duration of this phase depends on the value of the time constant T in (33) (which determines the change of the variable n with respect to time), as well as on parameters A, n_B and M. In fig. 24, this region is located near the point V_{max}. The arrows on the I_{inw} and I_{outw} curves show the development of the depolarization and repolarization processes. Fig. 24 and equation (32) lead to the conclusion that the greater T or A are, the longer the period of slow repolarization. With the decrease of the parameter n_B, the period of slow repolarization as well as the APD shortens. When $n \geq n_B$, the excess of the outward current over the inward current increases in time, causes a smooth increase in the repolarization rate, until its maximum value is reached. When the AP is close to V_{rest}, this excess decreases together with $f(V)$, and the AP slowly approaches V_{rest}. Parameter n_B also affects the APD restitution curve. The APD restitution curve (4) on Fig. 21 corresponds to $n_B = 0.507$.

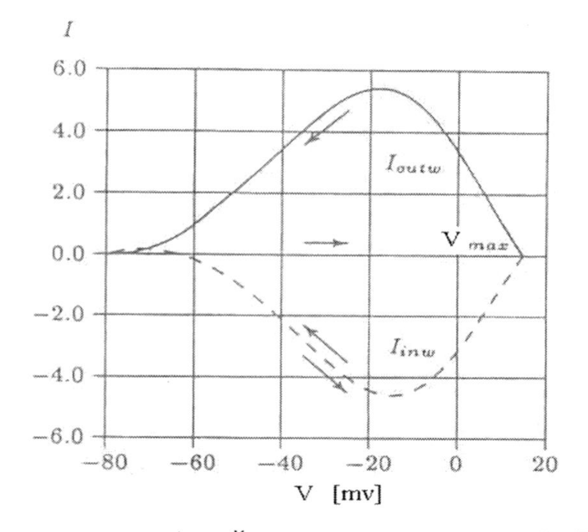

Fig. 24: I_{inw}=-$(Af(V) - V)$ and $I_{outw} = (n/n_B)^M f(V)$ currents for the Karma model. The arrow pointing from left to right designates the change of inward and outward currents in the depolarization phase of AP, and arrows pointing in the opposite direction show the changes of the same currents in the repolarization phase.

The AP reproduced by Karma Model is shown in fig. 25

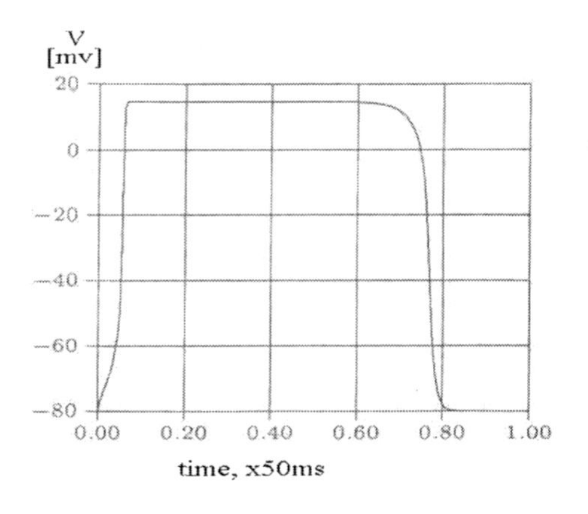

Fig. 25: AP generates by the Karma model.

5.5. Conclusion

We state that the existing simplified second-order models exhibit the following properties:

(1) All of them are based on the same assumptions and differ in the analytical approximations of the experimental dependencies of I_{inw} and I_{outw} on V, and in this way show how the slow variable affects the outward current.

(2) In comparison with the electrophysiological ionic models, they all produce decreased $([dV/dt])_{max}$ (due to the substitution $h = h_\infty$, $m = m_\infty$) and introduce time constants of slow variable, T, that do not depend on V. This independence leads to substantial errors in reproducing the APD restitution.
(3) In all simplified models considered here, the time-independent component of the outward current is neglected. This excludes the overshoot in generated AP. The VCD model is an exception, where overshoot can be obtained by changing the slope of the last segment of the piece-wise-linear approximation of current i_l.

Analysis of the Beeler-Reuter [9] and Luo and Rudy [10] ionic models shows that the introduced gate variable j in the sodium channel does not return to its steady state value ($j=1$) in the case of a short diastolic interval. That decreases the $(dV/dt)_{max}$ of the depolarization phase of the next AP and correspondingly decreases the conduction velocity. Second-order simplified models do not reproduce this phenomenon.

5.6. References

1. Paulsen, R.A., Jr., J.W. Clark, Jr., P.H. Murphy, and J.A. Burdine, *Sensitivity analysis and improved identification of a systemic arterial model.* IEEE Trans Biomed Eng, 1982. **29**(3): p. 164-77.
2. Tikhonov, A.N., *Sets of differential equations containing small parameters on derivatives.* Math. Collection, 1952. **31**(73): p. 575-586.
3. Krinsky, V.I. and Y.M. Kokos, *Analysis of the equations of excitable membranes. III. Membrane of the Purkinje fibre. Reduction of the noble equations to a second order system. Analysis of automation by the graphs of the zero-isoclines.* Biofizika, 1973. **18**(6): p. 1067-1073.
4. Van Der Pol, B. and J. Van Der Mark, *The Heartbeat Considered as Relaxation Oscillations, and an Electrical model of the Heart.* Archives Neerlandaises Physiologe De L'Homme et des Animaux, 1929. **XIV**: p. 418-443.
5. FitzHugh, R., *Mathematical Models of Excitation and Propagation in Nerve*, in *Biological Engineering*, H.P. Schwan, Editor. 1969, McGraw-Hill: New York. p. 1-85.
6. Nagumo, J., S. Arimoto, and S. Yoshizawa, *An Active Pulse Transmission Line Simulating Nerve Axon.* Proceedings of the IRE, 1962. **50**(10): p. 2061-2070.
7. Ivanitsky, G.R., V.I. Krinsky, and E.E. Selkov, *Mathematical Biophysics of Cell.* 1978, Moscow: Nauka Press.
8. Karpoukhin, M.G., B.Y. Kogan, and W.J. Karplus, *The Application of a Massively Parallel Computer to the Simulation of Electrical Wave Propagation Phenomena in the Heart Muscle Using Simplified Models.* Proceedings of the 28th Annual Hawaii International Conference on System Sciences, 1995: p. 112-122.
9. Beeler, G.W. and H. Reuter, *Reconstruction of the action potential of ventricular myocardial fibres.* J.Physiol.(Lond), 1977. **268**: p. 177-210.
10. Luo, C.H. and Y. Rudy, *A Model of the Ventricular Cardiac Action-Potential - Depolarization, Repolarization, and Their Interaction.* Circulation Research, 1991. **68**(6): p. 1501-1526.
11. Noble, D., *Modification of Hodgkin-Huxley Equations Applicable to Purkinje Fibre Action and Pace-Maker Potentials.* Journal of Physiology-London, 1962. **160**(2): p. 317-&.
12. Karma, A., *Spiral Breakup in Model-Equations of Action-Potential Propagation in Cardiac Tissue.* Physical Review Letters, 1993. **71**(7): p. 1103-1106.
13. Ermakova, E.A., A.V. Pertsov, and E.E. Shnol, *On the interaction of vortices in two-dimensional active media.* Physica D, 1989. **40**: p. 185-195.
14. Nadapurkar, P.I. and A.T. Winfree, *A computation study of twisted linked scroll waves in excitable media.* Physica D, 1987. **29**: p. 69-83.

15. Pertsov, A.V., R.N. Chramov, and A.V. Panfilov, *Sharp increase in refractory period induced by oxidation suppression in FitzHughNagumo model. New mechanism of anti-arrhythmic drug action.* Biofizika, 1981. **6**.

16. Pertsov, A.V. and A.V. Panfilov, *Spiral waves in active media. The reverberator in the FitzHugh-Agumo Model*, in *Autowave Processes in Systems with Diffusion*. 1981: Gor'kii. p. 77-91.

17. Winfree, A.T., *When Time Breaks Down: The Three-Dimensional Dynamics of Electrochemical Waves and Cardiac Arrhythmias.* 1987: Princeton Univ Press.

18. Winfree, A.T., *Electrical instability in cardiac muscle: phase singularities and rotors.* J Theor Biol, 1989. **138**(3): p. 353-405.

19. Kogan, B.Y., W.J. Karplus, B.S. Billet, A.T. Pang, H.S. Karagueuzian, and S. Khan, *The simplified Fitzhugh-Nagumo model with action potential duration restitution: effects on 2D-wave propagation.* Physica D, 1991. **50**: p. 327-340.

20. van Capelle, F.J. and D. Durrer, *Computer simulation of arrhythmias in a network of coupled excitable elements.* Circ Res, 1980. **47**(3): p. 454-66.

Chapter 6. Computer Implementation of Mathematical Models

Computer implementation of an AP mathematical model requires:

a. A well-defined statement of the problem for computer simulation
b. Selecting a computer architecture – a sequential or parallel
c. Choosing the most effective numerical algorithms for the problem under investigation
d. Investigating the possibility of utilizing standard (MATLAB, Mathematica, etc.) and specialized software (OXSOFT, Madonna, Visualization programs) packages
e. Providing programming tools for measuring the conduction velocity of the wavefront and representing the cell's state in time for chosen grid points in the spatial domain.

6.1. Numerical methods for solving ordinary differential equations

Consider the following nonlinear ordinary differential equation:

$$\frac{dy}{dt} = f(y,t), \qquad y = y_0 \;\; at \;\; t = 0. \tag{1}$$

Let us introduce a discrete-time variable, t. For simplicity, let us choose equal time steps denoted by the variable h:

$$h = (t_1 - t_0) = \cdots = (t_{K+1} - t_K) = \cdots,$$

where discrete samples of t are shown below on a continuous interval.

0 1 2 3 .... K K+1 t

Two types of numerical methods are used for solving (1) [1]:

- Single step or self originated methods (self-starting)
- Multi-steps or non-self originated methods (non-self starting)

6.1.1. Methods Based on Taylor Series Representation

For the first time step:

$$y_1(t_0 + h) = y_0 + y_0^{(1)}\frac{h}{1!} + y_0^{(2)}\frac{h^2}{2!} + y_0^{(3)}\frac{h^3}{3!} + \ldots y_0^{(n)}\frac{h^n}{n!} + \cdots, \tag{2}$$

where:

$$y_0^{(1)} = \frac{dy}{dt}\Big|_{at\; t=t_0, y=y_0} = f(y_0, t_0); \quad y_0^{(n)} \equiv \frac{d^n y}{dt^n}\Big|_{at\; t=t_0, y=y_0} = \frac{d^{n-1} f(y,t)}{dt^{n-1}}\Big|_{at\; t=t_0, y=y_0}.$$

For the $K+1^{st}$ step:

$$y_{K+1} = y_K + y_K^{(1)}\frac{h}{1!} + y_K^{(2)}\frac{h^2}{2!} + y_K^{(3)}\frac{h^3}{3!} + \ldots + y_K^{(n)}\frac{h^n}{n!} + \ldots, \tag{3}$$

B.Ja. Kogan, *Introduction to Computational Cardiology: Mathematical Modeling and Computer Simulation*, DOI 10.1007/978-0-387-76686-7_6,

where:

$$y_K^{(1)} = f(y_K, t_k)$$

$$y_K^{(2)} = \left(\frac{d}{dt} y^{(1)}\right)_{t=t_K and\ y=y_K} = \frac{d}{dt} f(y,t)| at\ t = t_K, y = y_K$$

$$y_K^{(n)} = \frac{d^{n-1}}{dt^{n-1}} f(y,t)| \ at\ t = t_K, y = y_K$$

This simple analysis shows that the solution of a first order differential equation requires n-1 differentiations of the given function $f(y, t)$.

Let us consider, as an example [2], the solution of the following linear ordinary equation:

$$\frac{d^2y}{dt^2} + 2\frac{dy}{dt} + y = \sin wt \text{ , with initial conditions } \dot{y}(0) = 2,\ y(0) = 3 \ .$$

In state variable form, this equation takes the form:

$$\frac{dy}{dt} = z \tag{4}$$

$$\frac{dz}{dt} = \sin wt - 2z - y; \quad z_0 = \left(\frac{dy}{dt}\right)_0 = 2,\ y_0 = 3. \tag{5}$$

Taylor series expansion for y and z at the point t_1 gives:

$$y_1 = y_0 + h\left(\frac{dy}{dt}\right)_0 + \frac{h^2}{2!}\left(\frac{d^2y}{dt^2}\right)_0 + \frac{h^3}{3!}\left(\frac{d^3y}{dt^3}\right)_0 + \cdots \tag{6}$$

$$z_1 = z_0 + h\left(\frac{dz}{dt}\right)_0 + \frac{h^2}{2!}\left(\frac{d^2z}{dt^2}\right)_0 + \frac{h^3}{3!}\left(\frac{d^3z}{dt^3}\right)_0 + \cdots . \tag{7}$$

The higher derivatives in equations (6)-(7) are determined using repeated differentiation of equations (4)-(5):

$$\frac{d^2y}{dt^2} = \frac{dz}{dt} = \sin wt - 2z - y \ , \ \frac{d^2z}{dt^2} = w\cos wt - 2\frac{dz}{dt} - \frac{dy}{dt},$$

$$\frac{d^3y}{dt^3} = \frac{d^2z}{dt^2} = w\cos wt - 2\frac{dz}{dt} - \frac{dy}{dt}, \text{ and } \frac{d^3z}{dt^3} = -w^2 \sin wt - 2\frac{d^2z}{dt^2} - \frac{d^2y}{dt^2} .$$

So, for w=1:

$$\left(\frac{dy}{dt}\right)_0 = 2,\ \left(\frac{dz}{dt}\right)_0 = -7,\ \left(\frac{d^2y}{dt^2}\right)_0 = -7,\ \left(\frac{d^2z}{dt^2}\right)_0 = 13,\ \left(\frac{d^3y}{dt^3}\right)_0 = 13, \text{ and}$$

$$\left(\frac{d^3z}{dt}\right)_0 = -19.$$

If $h = 10^{-2}$ sec:

$$y_1 = 3 + 10^{-2} * 2 - \left(\frac{10^{-4}}{2} * 7 \right) + \left(\frac{10^{-6}}{6} * 13 \right) + \cdots \tag{8}$$

$$z_1 = 2 - 10^{-2} * 7 + \left(\frac{10^{-4}}{2} * 13 \right) + \left(\frac{10^{-6}}{6} * 19 \right) + \cdots . \tag{9}$$

It can be easily observed that the higher order terms in equations (8)-(9) quickly become negligibly small with the choice of a sufficiently small time step h.

A major advantage of this technique is that it allows estimation of the maximum error on each integration step. A drawback of this technique is that it requires numerical differentiation operations, which introduce additional errors when $f(y, t)$ is a complex function or is defined by a lookup table.

6.1.2. Euler's Method

Consider the Taylor expansion at the point t_K

$$y_{K+1} = y_K + \frac{h}{1!} y_K^{(1)} + \frac{h^2}{2!} y_K^{(2)} + \frac{h^3}{3!} y_K^{(3)} + \cdots$$

Assume that all derivatives shown do exist. Now if $y_K^{(2)}$ is bounded and h is small, we may ignore all terms after the first two and obtain:

$$y_{K+1} \approx y_K + h y_K^{(1)} = y_K + hf(y_K, t_K); \qquad K = 0, 1, \ldots N-1.$$

In the above expression, $\approx$ means "approximately equal to." This numerical scheme is known as Euler's method.

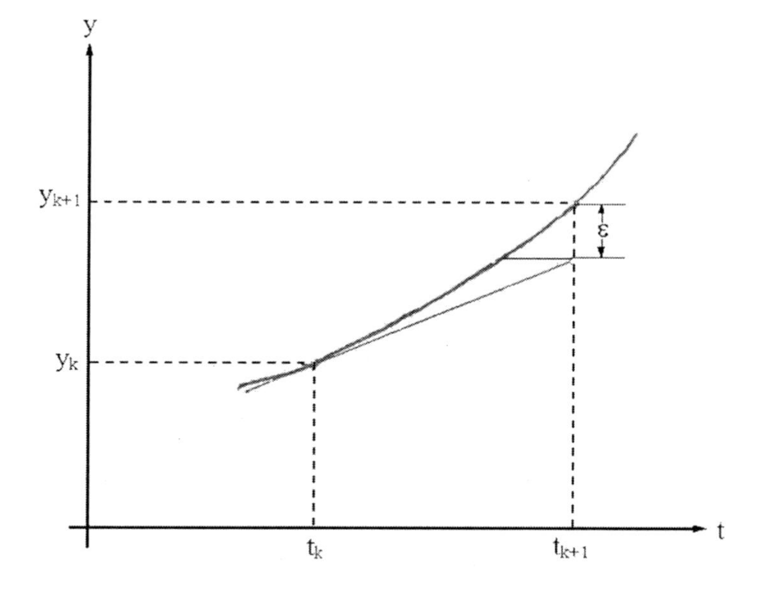

Fig. 1. Graphical representation of one step for Euler's method.

Geometrically, Euler's method consists of approximating the solution at t_{K+1} by following the tangent to the solution curve at point t_K (see Fig. 1).

If we denote $M = \max\left|y^{(2)}(t)\right|$, $0 \le t \le T$, the local truncation error of Euler's method can be expressed as:

$$L(h) \le \frac{h^2}{2} M = O(h^2)$$

The standard notation *O(h)* denotes the quantity that approaches zero at the same rate as h.

The global truncation error when $y^{(2)}(t)$ is bounded can be expressed as:

E(h) = O(h).

Therefore, the error decreases proportionally to the decrease in step size h.

In order to make the error tend to zero at a faster rate than h, other methods were developed. As an example, we examine the Heun or second-order Runge-Kutta method (RK2).

6.1.3. Second Order Runge-Kutta (RK2) Method

$$y_{K+1} = y_K + \frac{1}{2}\left[f(y_K, t_K) + f\left([y_K + hf(y_K t_K)], t_{K+1} \right) \right] h,$$

where $y_K + hf(y_K, t_K) = y^*_{K+1}$.

Denoting $k_0 = hf(y_K, t_K)$ and $k_1 = hf\{[y_K + hf(y_K, t_K)], t_{K+1}\}$, we finally obtain:

$$y_{K+1} = y_k + \tfrac{1}{2}(k_0 + k_1). \tag{10}$$

Here, we replaced $f(y_K, t_K)$ in Euler method by an average of the function f evaluated at the beginning t_k and the end t_{k+1} of time step h, as illustrated in fig. 2. The local truncation error is O(h^3). The global truncation error is E(h) = O(h^2).

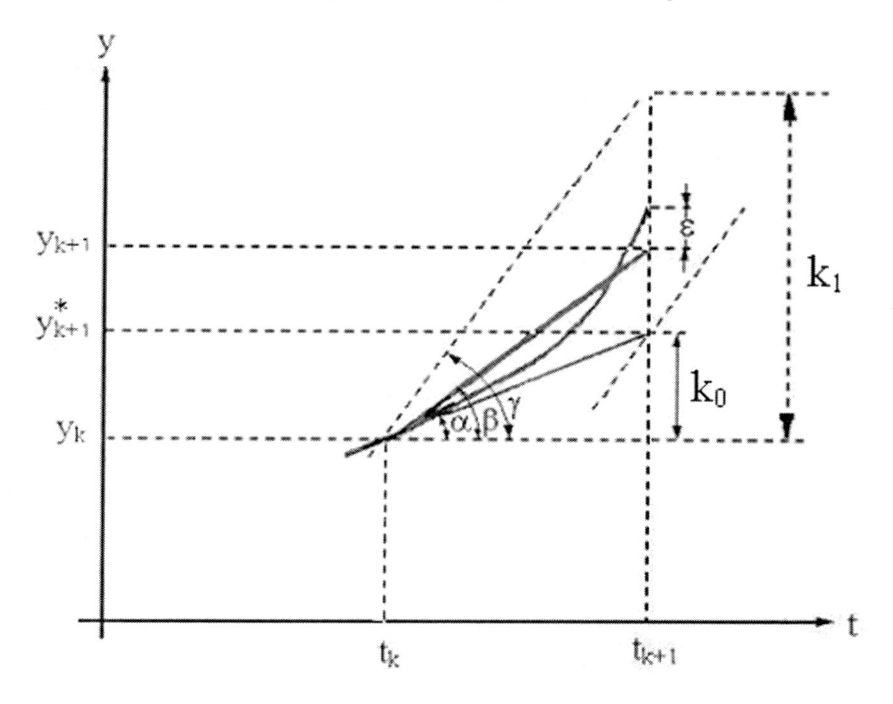

Fig. 2. Graphical representation of one step for the second order Runge-Kutta method.

6.2. Numerical solution of a system of ordinary differential equation

6.2.1. Euler's Method:

Let us consider the numerical solution of the following system of equations:

$$\dot{y}_1 = f_1(y_1, y_2, t) \tag{11}$$

$$\dot{y}_2 = f_2(y_1, y_2, t), \tag{12}$$

with the given initial conditions $y_1(0) = \hat{y}_1,\ y_2(0) = \hat{y}_2$

$$y_{1(K+1)} = y_{1K} + hf_1(t_K, y_{1K}, y_{2K})$$

$$y_{2(K+1)} = y_{2K} + hf_2(t_K, y_{1K}, y_{2K}),$$

where $K = 0, 1, 2, \ldots$

6.2.2. RK2 Method

For the same system of ordinary differential equations (ODEs):

$$y_{1K+1} = y_{1K} + \frac{1}{2}[k_0 + k_1]$$

$$y_{2K+1} = y_{2K} + \frac{1}{2}[l_0 + l_1],$$

where k_0, k_1, l_0, and l_1 are $k_0 = hf_1(t_K, y_{1K}, y_{2K})$, $k_1 = hf_1(t_{K+1}, y_{1K} + k_0, y_{2K} l_0)$, $l_0 = hf_2(t_K, y_{1K}, y_{2K})$, and $l_1 = hf_2(t_{K+1}, y_{1K} + k_0, y_{2K} l_0)$.

Using a fourth-order Runge-Kutta method for the system of simultaneous first-order differential equations, we obtain for equations (11)-(12):

$$y_{1K+1} = y_{1K} + \frac{1}{6}[k_0 + 2k_1 + 2k_2 + k_3] \tag{13a}$$

$$y_{2K+1} = y_{2K} + \frac{1}{6}[l_0 + 2l_1 + 2l_2 + l_3], \tag{13b}$$

where k_0, k_1, k_2, k_3, l_0, l_1, l_2, and l_3 are

$$k_0 = hf_1(t_K, y_{1K}, y_{2K}), k_1 = hf_1\left(t_K + \frac{1}{2}h, y_{1K} + \frac{1}{2}k_0, y_{2K} + \frac{1}{2}l_0\right),$$

$$k_2 = hf_1\left(t_K + \frac{1}{2}h, y_{1K} + \frac{1}{2}k_1, y_{2K} + \frac{1}{2}l_1\right), k_3 = hf_1(t_K + h, y_{1K} + k_2, y_{2K} + l_2),$$

$$l_0 = hf_2(t_K, y_{1K}, y_{2K}), l_1 = hf_2\left(t_K + \frac{1}{2}h, y_{1K} + \frac{1}{2}k_0, y_{2K} + \frac{1}{2}l_0\right),$$

$$l_2 = hf_2\left(t_K + \frac{1}{2}h, y_{1K} + \frac{1}{2}k_1, y_{2K} + \frac{1}{2}l_1\right),$$

$$l_3 = hf_2(t_K + h, y_{1K} + k_2, y_{2K} + l_2).$$

The local truncation error can be estimated as $L(h) = O(h^5)$ and the global error as $E(h)=O(h^4)$. See [1] and [2] for details.

6.2.3. The Ashour-Hanna Method

The classical explicit integration methods (such as Euler and Runge-Kutta) have very limited stability regions. Since the integration step size is restricted mainly by stability, rather than by truncation error considerations, these methods tend to become extremely inefficient [3]. More efficient implicit methods allowing the use of a much larger step size, such as the implicit Runge-Kutta and backward differentiation methods are used to perform the integration for moderately and mildly stiff problems. These methods, however, require more memory and more computation per time step.

S. S. Ashour and O. T. Hanna [4] proposed a new simple explicit method for the integration of mildly stiff ODEs. Let us consider the following system of ODEs:

$$y' = f(t, y),$$

with initial condition

$$y(t_0) = y_0.$$

Following the Ashour-Hanna (AH) method, we start from a specified or previously determined $y(t)$, and execute a single step using the first-order explicit Euler method:

$$y_{Euler}(t + h) = y(t) + hf[t, y(t)] \tag{14}$$

Then we carry out a single step using the second-order explicit Runge-Kutta-Trapezoidal method (RK2T) starting from the same $y(t)$:

$$y_{RK2T}(t+h) = y(t) + \left(h/2\right) \times \{f[t, y(t)] + f[t+h, y_{Euler}(t+h)]\}. \tag{15}$$

Finally, we average the Euler and RK2T values to obtain the new value as follows:

$$y(t+h) = \alpha y_{Euler}(t+h) + (1-\alpha) y_{RK2T}(t+h). \tag{16}$$

After substituting (14) and (15) in (16), we obtain :

$$y(t+h) = \alpha\{ y(t) + hf[t, y(t)]\} + (1-\alpha)\{ f[(t, y(t)] + f[t+h, y(t) + hf(t, y(t)]\} \tag{17}$$

where α is an averaging parameter, $0 < \alpha < 1$. When $\alpha = 0$, the AH algorithm is reduced to the second order RK2T; when $\alpha = 1$, it is reduced to the first-order Euler method. The parameter α is chosen so as to minimize spurious oscillations in the solution and maximize the permissible integration step size.

6.2.4. The Hybrid Integration Method

For sufficiently small membrane potential offsets, the rate constants $\alpha_{y_i}(V)$ and $\beta_{y_i}(V)$ remain essentially unchanged over the corresponding time interval, Δt. The approximate solution of the gate variable equations

$$\frac{dy_i}{dt} = \alpha_{y_i}(1 - y_i) - \beta_{y_i} y_i$$

or

$$\tau_{y_i}(V)\frac{dy_i}{dt} = y_{i_\infty}(V) - y_i, \tag{18}$$

can be written as a simple exponential of the form:

$$y_i = y_{i\infty} + [y_{i0} - y_{i\infty}]\, e^{-\frac{\Delta t}{\tau_{y_i}}}. \tag{19}$$

So, the numerical integration of (18) can be replaced by (19) for a given time increment Δt. An automatic procedure can be introduced to adjust the time increment Δt so that the membrane potential offset ΔV remains between required limits.

6.3. Model Implementation on Parallel Supercomputers

6.3.1. Mathematical Model for AP Generation and Propagation (Generalized Form)

$$\frac{\partial V}{\partial t} = \frac{1}{C}F(V, \vec{U}) + D\nabla^2 V^* + \frac{1}{C}I_{st}(t), \ \dim\vec{U} = N^* \tag{20}$$

---

* For non-uniform anisotropic tissue, the expression $D\nabla^2 V$ is replaced by $\nabla(D\nabla V)$. Here, ∇ is the gradient operator.

$$\frac{\partial \vec{U}}{\partial t} = \vec{f}\left(V, \vec{U}, \vec{\mu}, t\right), \ \dim \vec{\mu} = M, \ \vec{\mu} = \vec{\mu}(t). \tag{21}$$

Here:

- V is the cell membrane potential [mV]
- t is time [ms]
- D is the diffusion coefficient [cm^2/s]
- ∇^2 is the Laplacian operator
- $F(V, \vec{U})$ is a nonlinear function which represents the sum of the ionic currents [$\mu A/cm^2$]
- $I_{st}(t)$ is the external applied stimulus current [$\mu A/cm^2$]
- C is the membrane capacitance [$\mu F/cm^2$]
- $\vec{U}$ describes the dynamics of ionic channel permeabilities (gate variables) and ionic concentrations in intracellular compartments
- $\vec{\mu}$ is the vector of the parameters, which varies with time.

These equations are solved with the following initial and boundary conditions: $V(\vec{r},0), \vec{U}(\vec{r},0)$ and $\frac{\partial V}{\partial \vec{n}} = 0$, where $\vec{r}$ is the vector of space coordinates and $\vec{n}$ is the direction of the normal to the boundary G.

When D=0 (no diffusion), equations (20) and (21) are reduced to the so-called point model, which reproduces the AP generation by a single cell.

The simplest second-order FitzHugh-Nagumo point model has only one generalized gate variable (N=1 and M=0 in (20) and (21)) and three nonlinear functions of V, while the most sophisticated thirteenth-order Luo and Rudy II model [8, 9] includes 9 equations for ionic gate variables and 3 for $[Ca^{2+}]_i$ in intracellular compartments (N=12 in (20) and (21)) and 33 nonlinear functions.

6.3.2. Implementation on Parallel Computers

The objective of a computer simulation is to find the distribution of membrane potential in time and space for a 2D or 3D model of cardiac tissue. The cardiac tissue, assumed to be a uniform isotropic syncytium (continuous medium) is approximated by a grid of 256×256 nodes connected by coupling resistors. The operator splitting algorithm [10] allows adaptive time step integration. According to this method, the integration of (20) and (21) is split into two parts:

integration of the diffusion equation

$$\frac{\partial V}{\partial t} = D \nabla^2 V, \tag{22}$$

and integration of the point model equations

$$\frac{\partial V}{\partial t} = \frac{1}{C} F\left(V, \vec{U}\right) + \frac{1}{C} I_{st}(t) \tag{23}$$

$$\frac{\partial \vec{U}}{\partial t} = \vec{f}\left(V, \vec{U}, \vec{\mu}, t\right), \ \vec{\mu} = \vec{\mu}(t). \tag{21}$$

Equations (22) and (23) are solved in the following sequence of events in each time cycle. First, equation (22) is solved during time step $\Delta t/2$ with the given initial and boundary conditions. Then, equations (21) and (23) are solved with the given initial conditions for the variable $\vec{U}$. For the variable V, the initial condition is obtained by using the solution of (22) at the end of time $\Delta t/2$. Equations (21) and (23) are stiff ODEs; therefore, integration with an adaptive time step is used to decrease the overall simulation time. We used $\Delta t_1 = 0.05\Delta t$ to integrate the stiff part of the equations. Finally, the values of V, obtained at the end of Δt, are used as initial conditions for the next integration step of the diffusion equation (22), during another time step $\Delta t/2$. This completes the first cycle. The data from the Laplacian calculations are exchanged between all processors twice during each time step Δt. Subsequently, the initial conditions for the next solution of equation (22) are taken from the solution at the end of the previous cycle.

The grid is divided between "n" available parallel processor units (PU) so that each processor is solving equations (21), (22) and (23) for $(256\times256/n)$ nodes. The geometrical apportionment of the grid nodes to the processors can be accomplished in different ways. The optimum one results in an equal computational load for all PUs and a minimum exchange of data between the PUs. We elected to divide the 2D tissue into parallel strips.

The explicit Euler numerical integration algorithm is used for equation (22) where the fixed time step $\Delta t/2$ = 0.1ms. Well-known implicit integration methods (such as alternative direction) are difficult to implement for parallel computations. Equations (21) and (23) are solved by the explicit Euler method with the exception of the equations for the gate variables with time constants $\tau_k \sim \Delta t_1$. In the Luo-Rudy II model in particular, only one time constant is encountered which is comparable with the smallest chosen time step, Δt_1. This time constant is associated with the gating variable m of the sodium channel. The solution for this gate variable is obtained by the so-called hybrid method proposed in [5] and [6] and analyzed in [7]. For this application, the hybrid method provides better stability then the Euler method.

6.4. Dimensionless Form of Equations for Wave Propagation

$$C\frac{dV}{dt} = \alpha_x \frac{\partial^2 V}{\partial x^2} + \alpha_y \frac{\partial^2 V}{\partial y^2} - \sum I_{ion_i} + I_{st}. \tag{24}$$

Here:

- V is membrane potential [mV]
- C is membrane capacitance [μF/cm2]
- t is time [ms]
- $\alpha_x = a/2R_i$ is coupling conductivity [1/kΩ]
- a is the cell radius equal to 5 - 10 μm or (5-10)10^{-4}cm
- R_i is the specific resistance of intracellular liquid, equals 0.2 kΩ-cm.

For the gating variables:

$$\tau_{y_i}(V)\frac{\partial y_i}{\partial t} = y_{i_\infty}(V) - y_{i,} \tag{25}$$

where y_i represents the gating variables (i.e. *m, h, j, n,* etc.).

Usually

$$I_{ion_i} = \bar{g}_i f_i(V, y) \text{ or } \bar{g}_i f_i(V)\left[\frac{\mu A}{cm^2}\right]. \tag{26}$$

Let us multiply both sides of (24) by the membrane resistance at equilibrium or rest potential, R_m, [kΩ-cm^2] and introduce the anisotropy ratio $A = \alpha_y / \alpha_x$. We then obtain:

$$R_m C\frac{\partial V}{\partial t} = \alpha_x R_m\left(\frac{\partial^2 V}{\partial x^2} + A\frac{\partial^2 V}{\partial y^2}\right) - R_m\sum_1^h (I_{ion})_i + R_m I_{st},$$

(27)

We designate:

$R_m C = \tau_f$

$R_m \alpha_x = \lambda_x^2$

λ_x - is the length constant.

Introducing dimensionless values:

- time $\qquad \bar{t} = t/\tau_f$
- space $\qquad \bar{x} = x/\lambda_x$ and $\bar{y} = y/(A\lambda_x)$
- potential $\qquad \bar{V} = |V|/|V_{max}|$
- currents $\qquad \bar{I}_i = I_i / I_{max}$.

We obtain finally for (27)

$$\frac{\partial \bar{V}}{\partial t} = \left(\frac{\partial^2 \bar{V}}{\partial \bar{x}^2} + \frac{\partial^2 \bar{V}}{\partial \bar{y}^2}\right) + \sum_i^h (\bar{I}_{ion})_i. \tag{28}$$

For (25), we obtain:

$$\frac{\tau_{y_i}}{\tau_f}\frac{\partial y_i}{\partial \bar{t}} = y_{i\infty}(\bar{V}) - y_i,$$

with the ratio $\tau_f / \tau_{y_i} = \varepsilon_i$; when $\tau_{y_i} >> \tau_f$, ε_i *are* the small parameters

6.5. Determination of Wave Front Velocity in a 2D Model of Tissue

Wavefront conduction velocity is defined as velocity measured along the normal at each given point on the front.

The wavefront appears at a point (i, j) (a node of the grid) when the following conditions are satisfied:

$$\left(\frac{dV}{dt}\right)_{ij} > 0$$

$$V_{ij} \geq V_{front}.$$

Assume that a wavefront has passed through the grid nodes (i, j), $(i+1, j)$ and $(i, j+1)$ at times t_1, t_2 and t_3 respectively (see fig. 3). Then, the velocity of wavefront propagation at point (i,j) can be computed as:

$$\theta = \frac{|AB|}{t_2 - t_1} = \frac{\Delta x \cos\alpha}{t_2 - t_1} \text{ and } \theta = \frac{|AC|}{t_3 - t_1} = \frac{\Delta y \sin\alpha}{t_3 - t_1}.$$

Therefore

$$\frac{\Delta x \cos\alpha}{t_2 - t_1} = \frac{\Delta y \sin\alpha}{t_3 - t_1}$$

and

$$\tan\alpha = \frac{t_3 - t_1}{t_2 - t}\frac{\Delta x}{\Delta y}.$$

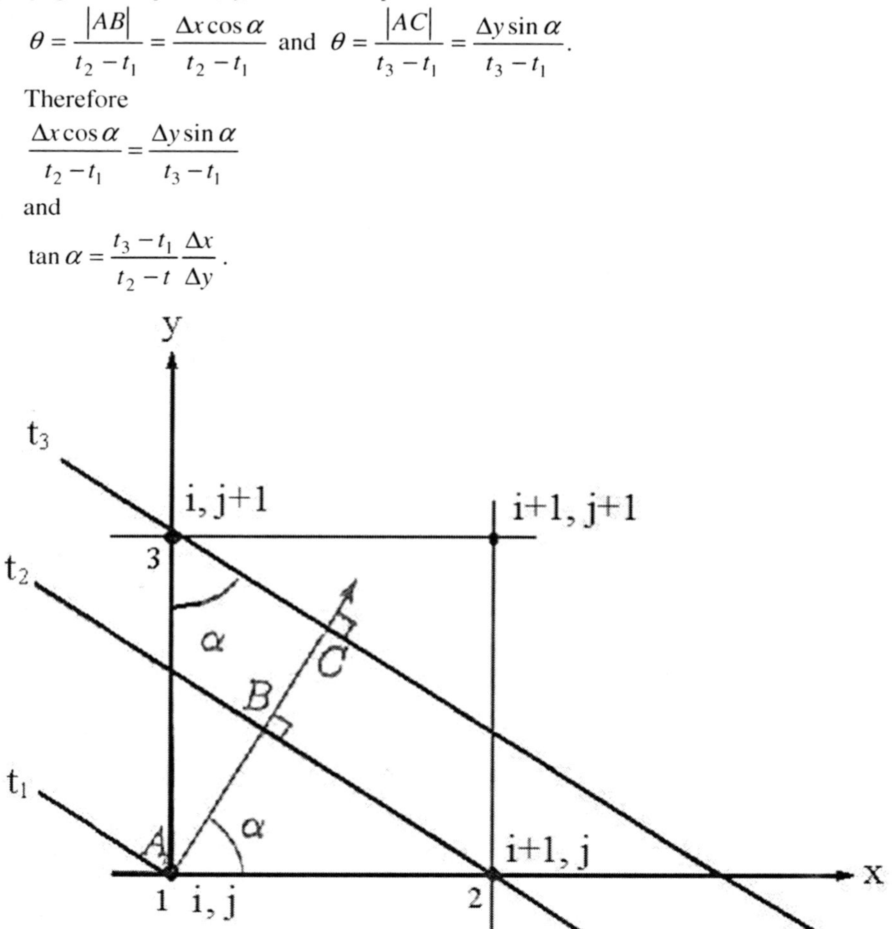

Fig. 3. Graphical method for determining wavefront conduction velocity.

We propose another approach. During the simulation, we can store the times (t_{ij}, $t_{i+1,j}$) when the wavefront visited the nodes (i, j) and (i+1, j) in memory. Thus, the projection of conduction velocity in the x direction is

$$(\theta_x)_{ij} = \frac{\Delta x}{t_{i+1,j} - t_{ij}}.$$

Taking into consideration that the direction of normal to the wavefront coincides with the direction of gradient of V at that point, we find that

$$\cot \alpha = \left(\frac{\partial V / \partial x}{\partial V / \partial y} \right)_{ij}.$$

The components of ∇V are well known from Laplacian calculations, so it is now possible to obtain:

$$\theta_{x_{ij}} = \theta_{n_{ij}} \cos \alpha.$$

and

$$\theta_{n_{ij}} = \frac{\theta_{x_{ij}}}{\cos \alpha} = \theta_{x_{ij}} \sqrt{1 + \tan^2 \alpha}.$$

6.6. Appendix: Stability Conditions for Parabolic Partial Differential Equation (PDE) Solutions

The stability of the solution is understood here as the ability of the solution to converge more accurately when the time and space steps (of the chosen numerical algorithm) are decreasing. Such stability is reached when special relationships between time and space steps are satisfied.

For 1-D parabolic problems:

$$D \frac{\partial^2 V}{\partial x^2} = \frac{\partial V}{\partial t}.$$

The stability condition for this case is: $\Delta t \le \frac{\Delta x^2}{2D}$. For our problems, usually D = 0.001 [cm^2/ms].

For 2-D parabolic problems:

$$D\left(\frac{\partial^2 V}{\partial x^2} + \frac{\partial^2 V}{\partial y^2}\right) = \frac{\partial V}{\partial t},$$

with the stability condition given by $\Delta t \le \frac{\Delta x^2}{4D}$.

A.Winfree recommended choosing Δx based on the following consideration: $\sqrt{\frac{DT_r}{\Delta x^2}} > 1$ [11]. In this equation T_r denotes the rise time in the AP depolarization phase (usually $T_r \approx 3ms$).

Implicit numerical algorithms do not require these limitations on time and space steps, but lead to the iterative solutions of large matrices when nonlinear parabolic systems have to be solved.

6.7. References

1. Hamming, R.W., *Numerical Methods for Scientists and Engineers*. 1962, Columbus: McGraw-Hill.
2. Bekey, G.A. and W.J. Karplus, *Hybrid Computation*. 1968, New York: Wiley Publishing.
3. Aiken, R.C., ed. *Stiff Computation*. 1985, Oxford University Press: Oxford.
4. Ashour, S.S. and O.T. Hanna, *A new very simple explicit method for the integration of mildly stiff ordinary differential equations.* Comp Chem Eng, 1990. **14**(267-272).
5. Moore, J.W. and F. Ramon, *On numerical integration of the Hodgkin and Huxley equations for a membrane action potential.* J Theor Biol, 1974. **45**: 249-273.
6. Rush, S. and H. Larsen, *A practical algorithm for solving dynamic membrane equations.* IEEE Trans Biomed Eng, 1978. **25**: 389-392.
7. Victorri, B., A. Vinet, F.A. Roberge, and J.P. Drouhard, *Numerical integration in the reconstruction of cardiac action potentials using Hodgkin-Huxley-type models.* Comput Biomed Res, 1985. **18**: 10-23.
8. Luo, C.H. and Y. Rudy, *A dynamic model of the cardiac ventricular action potential. I. Simulations of ionic currents and concentration changes.* Circ Res, 1994. **74**: 1071-1096.
9. Luo, C.H. and Y. Rudy, *A dynamic model of the cardiac ventricular action potential. II. Afterdepolarizations, triggered activity, and potentiation.* Circ Res, 1994. **74**: 1097-1113.
10. Zeng, J., K.R. Laurita, D.S. Rosenbaum, and Y. Rudy, *Two components of the delayed rectifier K+ current in ventricular myocytes of the guinea pig type. Theoretical formulation and their role in repolarization.* Circ Res, 1995. **77**: 140-152.
11. Strang, G., *On the construction and comparison of difference schemes.* SIAM J Numer Anal, 1968. **5**: 506-517.
12. Winfree, A.T., *Heart muscle as a reaction-diffusion medium: the roles of electrical potential diffusion, activation front curvature, and anisotropy.* Int J Bif Chaos, 1997. **7**: 487-526.

Chapter 7. Excitation-Propagation in One Dimensional Fibers

The study of pulse propagation in one-dimensional (1D) fiber is of prime interest for the propagation through nerve fibers. For cardiac tissue, which is predominately 2D and 3D, this study presents chiefly methodological value. The exception is a type of the atrium flutter and observed circulation of excitation in atrium around vena cava.

It is worthwhile to consider two major cases: propagation along the fiber with open ends and propagation in a ring-shaped 1D fiber. For the first case we will consider the propagation of a solitary pulse and pulse sequences generated at one of the open ends.

The study of excitation wave propagation in a ring of cardiac tissue is a subject of significant practical and theoretical importance [1-3]. Methodologically it allows us to investigate the behavior of the cell in the fiber under different pacing rates by only changing the equivalent ring length. The study of excitation pulse propagation in a ring facilitates an understanding of mechanisms of many life-threatening cardiac tachyarrhythmias.

The physiological studies of excitation wave circulation in certain ring-shaped preparations of atrial tissue [4,6] show unstable propagation with irregular oscillations of the action potential duration (APD), period of circulation, and conduction velocity (CV).

7.1. Characteristics of excitation-propagation in a fiber

7.1.1. The cable theory

One-dimensional fiber is considered here as an extended cylindrical cell membrane (see Fig. 1.) This simplification is correct only when the gap junction resistance is negligible.

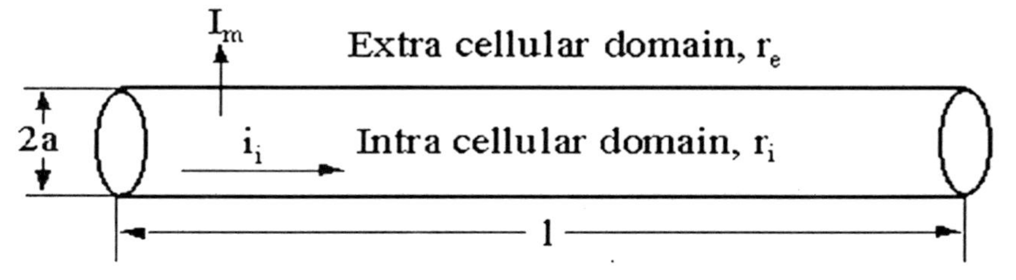

Fig. 1. Schematic representation of an extended excitable cell.

For fiber of such geometry and with intracellular resistance $r_i >> r_e$ (mono domain approach), radius a and length l:

$$\frac{\partial V_m}{\partial x} = -r_i i_i , \tag{1}$$

B.Ja. Kogan, *Introduction to Computational Cardiology: Mathematical Modeling and Computer Simulation*, DOI 10.1007/978-0-387-76686-7_7,

where: $r_i = R_i S_v$, R_i is specific resistance of intracellular liquid [Ω cm]; S_V denotes the surface to volume ratio $S_v = \frac{2\pi a l}{\pi a^2 l} = \frac{2}{a}$, V_m is difference of potential between intra- and extra-cellular membrane surfaces.
The full membrane current, I_m, is defined as:

$$I_m = -\frac{\partial i_i}{\partial x} \tag{2}$$

Eliminating i_i, in (1) by using (2) we obtain

$$\frac{\partial^2 V}{\partial x^2} = -r_i \frac{\partial i_i}{\partial x} = r_i I_m \tag{3}$$

Finally replacing I_m by the sum of membrane currents (see (11) chapter 3):

$$\frac{1}{r_i}\frac{\partial^2 V}{\partial x^2} = C_m \frac{\partial V}{\partial t} + \sum I_S + I_{St} \tag{4}$$

For the fiber considered as a continuous membrane of finite length L, (4) is valid and the initial and boundary conditions are:

$$V_m(0,x) = V_{m,rest}; \quad \frac{\partial V_m(t,0)}{\partial x} = \frac{\partial V_m(t,L)}{\partial x} = 0$$

When $I_{St} < I_{Th}$ in (4), the cell is not excited and applied stimulus propagates through the passive equivalent electrical circuit. This called passive propagation. In the opposite case, we deal with active propagation. The discrete equivalent circuit diagrams for active and passive propagation are shown in fig. 2a,b. These figures reflect the macro approach used for investigation of active (excitation) and passive wave propagation in one-dimensional tissue.

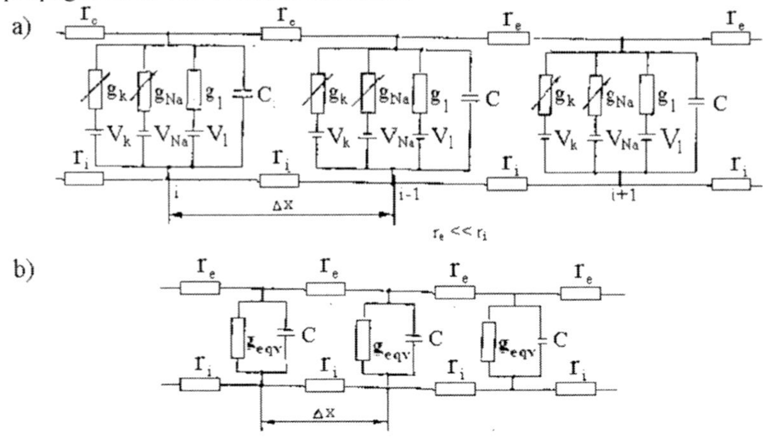

Fig. 2. Equivalent circuit diagrams for: a). Active propagation; b). Passive propagation

In cases where the gap junction resistance cannot be neglected relative to intracellular resistance, r_i, the discrete micro approach has to be used. One version of this approach, proposed in [7], is shown in Fig. 3.

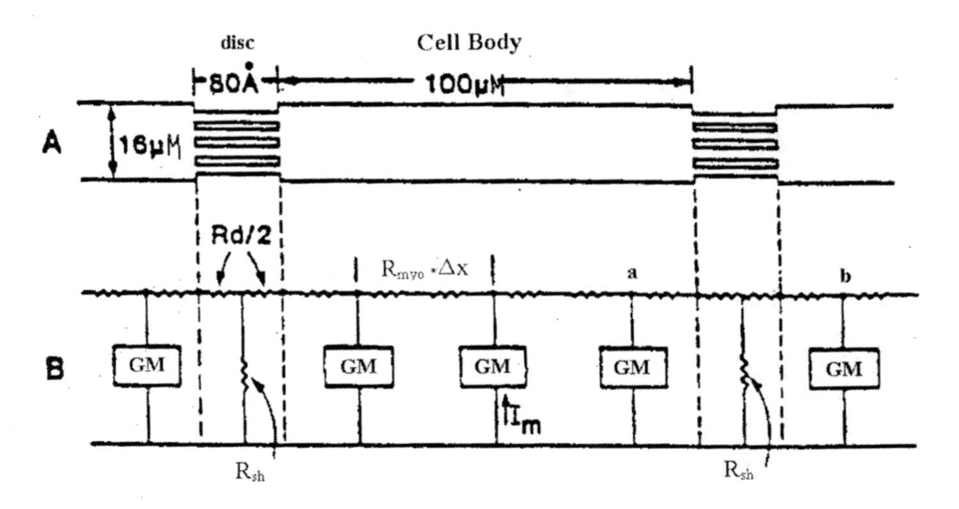

Fig. 3. A) Discrete cable model of cylindrical cardiac cells, each 100 μm in length and 16 μm in diameter, interconnected by an intercalated disk structure that contains intercellular bridges (connexons). B) Core conductor network with 3 generalized AP model (G-M) of membrane patches per cell, and a T network representing the intercalated disk between cells. R_d, disk resistance; R_{myo} = r_i, myoplasmic resistivity; and R_{sh}, leakage resistance to extra cellular space.

7.1.2. Passive propagation

For simplicity, we consider a semi-infinite fiber, initially at rest potential. A small potential $E_l < E_{th}$ is then applied at one end.

The equation (4) for this case may be rewritten as follows:

$$\frac{1}{r_i}\frac{\partial^2 E}{\partial x^2} = C_m \frac{\partial E}{\partial t} + E g_{eqv} \tag{5}$$

$$E(0,0) = E_1;\ \ E(t,\infty) = 0$$

- $g_{eqv} = \frac{1}{R_m}$, R_m is a membrane resistance at rest potential.
- E is depolarized increment of the membrane voltage smaller than $E_{th} = V_{m,th} - V_{m,rest}$

The equation (5) has a solution:

$$E(x,t) = \frac{E_1}{2}\left[-e^{\left(-x\sqrt{r_i g_{eqv}}\right)} erfc\left(\frac{x}{2}\sqrt{\frac{C_m r_i}{t}} - \sqrt{\frac{t\, g_{eqv}}{C_m}} \right)\right.$$

$$\left. +e^{-x\sqrt{r_i\ \ g_{eqv}}} erfc\left(\frac{x}{2}\sqrt{\frac{C_m r_i}{t}} + \sqrt{\frac{t\, g_{eqv}}{C_m}} \right)\right]$$

or

$$E(x,t)=\frac{E_1}{2}e^{-x\sqrt{g_{eqv}r_i}}\left[-\mathrm{erfc}\left(\frac{x}{2}\sqrt{\frac{C_m r_i}{t}}-\sqrt{\frac{tg_{eqv}}{C_m}}\right)\right.$$
$$\left.+\mathrm{erfc}\left(\frac{x}{2}\sqrt{\frac{C_m r_i}{t}}+\sqrt{\frac{tg_{eqv}}{C_m}}\right)\right] \tag{6}$$

Properties of the error function:

1. Definition: $\mathrm{erf}(x)=\frac{2}{\sqrt{\pi}}\int_0^x e^{-t^2}dt$

   $\mathrm{erfc}(x)=1-\mathrm{erf}(x)$
2. $\mathrm{erf}(0)=0;\ \mathrm{erf}(\infty)=1$
3. $\mathrm{erf}(x)=-\mathrm{erf}(-x)$

Let us find the spacial distribution of E, when $t \rightarrow \infty$ (stationary state). From (6) and properties 2, and 3, it follows:

$$E(x,\infty)=E_1 e^{-x\sqrt{g_{eqv}r_i}} \tag{7}$$

Defining $\frac{1}{\sqrt{g_{eqv}r_i}}=\lambda$ in (7), we finally obtain:

$$E(x,\infty)=E_1 e^{-\frac{x}{\lambda}}$$

λ is called the length constant and determines the decay of a local potential propagation in space. Namely, when $x = \lambda,\ E=\frac{E_1}{e}$

For example:

Calculate the value of λ if the cell parameters are:
$a=8-10\mu M$
R_i = 0. 2 [kΩ cm]
R_m= 6.25 [kΩ cm^2]

7.1.3. Active propagation (mono-domain approach)

Let us consider the equation (4). After division by C_m, we obtain:

$$\frac{1}{r_i C_m}\frac{\partial^2 V_m}{\partial x^2}=\frac{\partial V_m}{\partial t}+\frac{1}{C_m}\sum I_S+\frac{1}{C_m}I_{st} \tag{8}$$

Here $\frac{1}{r_i C_m}=D$ is called the diffusion coefficient and has the dimension $\left[\frac{cm^2}{s}\right]$

After multiplying the denominator of expression for D by $1= R_m\, g_{eqv}$, we obtain:

$$D = \frac{1}{r_i g_{eqv}(R_m C_m)} = \frac{\lambda^2}{\tau_m} \tag{9}$$

$$g_{eqv} = \frac{1}{R_m}; \ \tau_m = R_m C_m$$

It follows from (9) that diffusion coefficient expresses the passive properties of a membrane.

7.2. Bidomain approach

In this case we cannot neglect the value of r_e in comparison to $r_i = 1/g_i$. Moreover let us consider for generality that conductivities of extra- and intra-cellular liquid show anisotropy along the fiber axis x. Then, $g_e(x)$ and $g_i(x)$ are given functions of x.

By definition: $V_m = V_i - V_e$. According the Ohm's law the decrease in potential per unit length along the intra- or extra-cellular paths equals axial current times the resistance:

$$\frac{\partial V_i}{\partial x} = -r_i i_i; \ \frac{\partial V_e}{\partial x} = -r_e i_e \tag{10}$$

The loss of longitudinal current (per unit length) must precisely equal the transmembrane current according the Kirchoff's law. Indeed:

$$\frac{\partial i_i}{\partial x} = -I_m \text{ and } \frac{\partial i_e}{\partial x} = I_m + i_a \tag{11}$$

Here: $I_m = \left(C_m \frac{\partial V_m}{\partial t} + \sum I_S + I_{St} \right)$ is the transmembrane current measured in $\mu A/cm^2$, i_a is an external current applied to external domain and resembling defibrillation shock applied to tissues of higher dimensions.

Using equations (10) and (11) and expressions for V_m and I_m we obtain two basic equations for bidomain one-dimensional tissue representation:

$$\frac{\partial V_m}{\partial t} = \frac{1}{C_m}\left(\frac{\partial}{\partial x}\left(\frac{1}{r_i(x)} \frac{\partial V_m}{\partial x} \right) + \frac{\partial}{\partial x}\left(\frac{1}{r_i(x)} \frac{\partial V_e}{\partial x} \right) \right) - \frac{1}{C_m}(I_{ion} + I_{stim}) \tag{12}$$

and

$$\frac{\partial}{\partial x}\left(\left(\frac{1}{r_i(x)} + \frac{1}{r_e(x)} \right) \frac{\partial V_e}{\partial x} \right) = -\frac{\partial}{\partial x}\left(\frac{1}{r_i(x)} \frac{\partial V_m}{\partial x} \right) - I_a \tag{13}$$

Thus, the bi-domain approach in contrast to mono-domain requires solving a system of PDE: one parabolic equation (12), which describes the V_m propagation initiated by I_m and distribution of V_e obtained in the previous time step as solution of elliptic PDE (13) with appropriate boundary conditions and forcing function.

If r_i and r_e do not change along the fiber axis x, equations (12) and (13) may be simplified:

$$\frac{\partial V_m}{\partial t} = \frac{1}{C_m r_i}\left(\frac{\partial^2 V_m}{\partial x^2} + \frac{\partial^2 V_e}{\partial x^2} \right) - \frac{1}{C_m}\left(\sum I_S + I_{St} \right)$$

and

$$\left(\frac{1}{r_i}+\frac{1}{r_e}\right)\frac{\partial^2 V_e}{\partial x^2}=-\frac{1}{r_i}\frac{\partial^2 V_m}{\partial x^2}-I_a$$

These two equations can be reduced to one if we substitute the expression for $\frac{\partial^2 V_e}{\partial x^2}$ obtained from the last equation into the previous one. As a result, we get:

$$\frac{\partial V_m}{\partial t}=\frac{1}{C_m r_i}\frac{\partial^2 V_m}{\partial x^2}\left(\frac{1}{\frac{r_e}{r_i}+1}\right)-\frac{1}{C_m}\left(\sum I_S+I_{St}+I_a\frac{1}{1+\frac{r_i}{r_e}}\right) \tag{14}$$

Equation (14) reflects the behavior of 1D fiber with homogenous resistance distribution in *x*-direction for both domains. The bi-domain approach is expressed in decreasing the effective diffusion coefficient and presence of additional stimulus $I_a\frac{1}{1+\frac{r_i}{r_e}}$, originally applied to extra-cellular domain.

Indeed, let us denote the diffusion coefficient for bi-domain case as D_b, and ratio $\frac{r_e}{r_i}=\alpha$. Then, $D_b=D/\alpha+1$. Because $\alpha>0$, $D_b<D$. Additional stimulus $(I_{St})_{Add}=I_a\frac{1}{1+\frac{1}{\alpha}}$ will be smaller than I_a In the limit where $\alpha\to 0$ (14) smoothly becomes the equation for mono-domain approach.

For passive propagation:

$$D_b=\frac{\lambda_b{}^2}{\tau_m},\ \lambda_b{}^2=\frac{1}{(r_i+r_e)g_{eqv}}.\ \text{Thus, } D_b<D \text{ and } \lambda_b<\lambda$$

7.2.1. Velocity of propagation, θ

Physiological observation show that $\theta\approx\sqrt{D}$. For stationary propagation, θ is constant and it is possible to substitute $\zeta=x+\theta t$ in (8), which transforms the partial differential equation to an ordinary one:

$$D\frac{d^2V_m}{d\zeta^2}=\theta\frac{dV_m}{d\zeta}+\frac{\sum I_S+I_{St}}{C} \tag{15}$$

Here, θ is an unknown parameter – velocity of stationary propagation.

There exists an efficient numerical method [9, 12] to find θ. This method consists of multiple numerical integrations of the system (8) for constant initial conditions chosen near the initial stationary state, but each time for new values of θ.

Depending on the value of θ, the solution $V(\zeta)$ when $\zeta\to\infty$ will tend to either $+\infty$ or $-\infty$. Organizing computational process so as to find two close values θ_+ and θ_-

for which the solutions diverge to $+\infty$ and $-\infty$ respectively, one can get the value of stationary velocity in the form: $\theta = \dfrac{\theta_+ + \theta_-}{2}$.

The approximate computational method for calculation the variable propagation velocity in 1D and 2D tissue models is discussed below.

7.3. Excitation propagation in 1D fiber model

7.3.1. Propagation of the solitary pulse along a 1D fiber

As shown previously (see chapter 6), the mathematical model for excitation wave propagation in 1D tissue can be presented in either dimensional or dimensionless forms:

1. Dimensional form: for mono-domain approach:

$$C_m \frac{\partial V_m}{\partial t} = \frac{1}{r_i}\frac{\partial^2 V_m}{\partial x^2} - \sum_{i=1}^{N}(I_{ion})_i + I_{st} \tag{16}$$

$$\tau_{y_k}(V_m)\frac{\partial y_k}{\partial t} = y_{k\infty}(V_m) - y_k$$

Here: $\sum_{i=1}^{N}(I_{ion})_i = \sum (I_{ion})_{outw} - \sum (I_{ion})_{inw}$; $(I_{ion})_i = \overline{g}_i f_i(V_m, \prod y_k)(V_m - V_i)$

Initial conditions: $V_m(x,0)$, $y_k(x,0)$. Boundary conditions: $\left.\dfrac{\partial V_m}{\partial x}\right|_{x=-\infty} = \left.\dfrac{\partial V_m}{\partial x}\right|_{x=\infty}$

2. Dimensionless form:

$$\frac{\partial \overline{V_m}}{\partial \overline{t}} = \frac{\partial^2 \overline{V_m}}{\partial \overline{x}^2} + \sum (\overline{I}_{ion})_{inw} - \sum (\overline{I}_{ion})_{outw} + \overline{I_{st}} \tag{17}$$

$$\frac{\partial y_k}{\partial \overline{t}} = \varepsilon_k(\overline{V_m})[y_{k\infty}(\overline{V}_m) - y_k]$$

Here: $\varepsilon_k(\overline{V}_m) = \dfrac{\tau_f}{\tau_{y_k}(\overline{V}_m)}$ are the small parameters. The initial and boundary conditions are the same, but are expressed through dimensionless variables $\overline{x}$ and $\overline{V}_m$.

The main task for a computer simulation is to find the shape and velocity of the propagated pulse. The solution of that problem for a general case is rather difficult. It can be simplified if we consider the case when pulse is propagated with a constant velocity, θ.

For this particular case, using the substitution: $\xi = x + \theta t$ or $\overline{\xi} = \overline{x} + \overline{\theta t}$, we transfer our problem from the solution of PDEs to ODEs. Indeed, for dimensional case we obtain:

$$C\theta \frac{dV_m}{d\xi} = \alpha_x \frac{d^2 V_m}{d\xi^2} - \sum I_{ion} + I_{st} \tag{18}$$

$$\theta \tau_{y_k}(V) \frac{dy_k}{d\xi} = y_{k\infty}(V) - y_k \quad ; \quad \frac{1}{\tau_{y_k}(V)} = \varepsilon_k(V)$$

Using the H-H method described in paragraph 7.1.1, it is possible to find the steady state value of θ, which corresponds to the chosen parameters. Moreover, it is possible to find how θ depends on some parameters in equation (18).

For example, the dependence $\theta = f(\varepsilon)$ is shown in fig. 4 for the FH-N model reduced to dimensionless form.

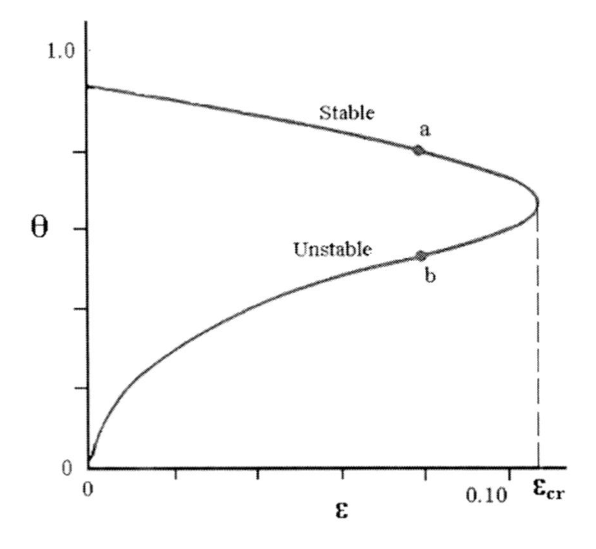

Fig. 4. Impulse velocity θ for stationary propagation vs. a small parameter ε [13] in FH-N model

The cell's outward current grows as ε increases. This leads to APD shortening and to the decrease of AP amplitude and $(\frac{dV_m}{dt})_{\max}$ values.

If ε is increased above ε_{cr}, no stationary pulse propagation is possible and conduction is blocked.

7.4. Propagation of pulse sequences

A pulse sequence occurs, for example, under periodic stimulation of a long fiber from one of its ends. For a given period of stimulation, one can observe the influence of the preceding pulse on the propagation of the subsequent one.

Three different cases can be considered:

1. Period of stimulation is long enough for recovery process to reach steady state (long diastolic interval).
2. Period of stimulation provides comparatively short DI.
3. Very short period of stimulation compared with full APD.

The tendency for equalization of the period of pulse sequence is observed at some distance from the point of stimulation. The velocity of pulse propagation decreases with decrease in period between pulses (see fig. 5.).

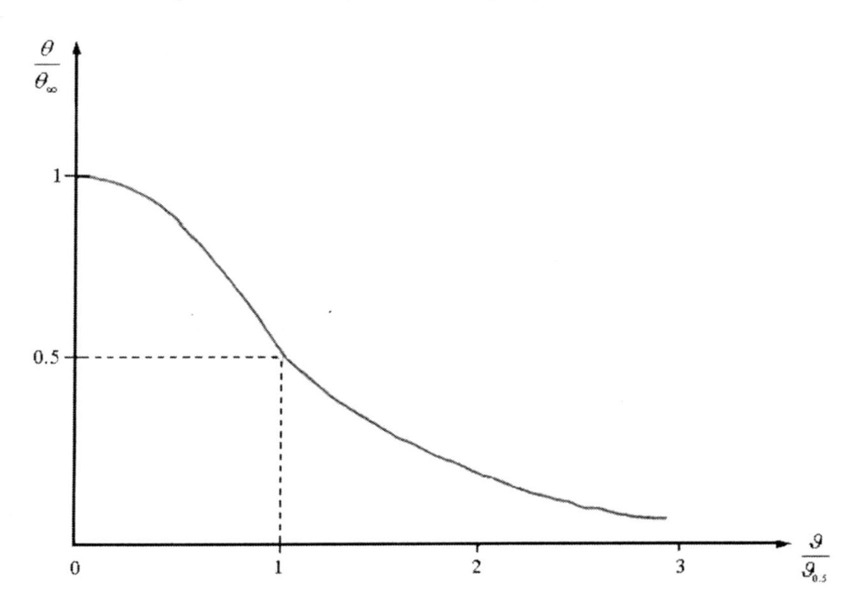

Fig. 5. Here $\vartheta = \frac{1}{T}$ - frequency of a pulses sequence, T- period of pulses sequence, $\vartheta_{0.5}$ is the frequency of a pulses sequence for which $\theta = 0.5\theta_\infty$ and θ_∞ is the velocity of propagation when $\vartheta = 0$ (single pulse propagation)

7.5. Propagation of excitation wave in model of a ring-shaped tissue

7.5.1. Initiation of pulse propagation in a ring-shaped tissue

Circulation of a pulse around a ring-shaped excitable tissue can be started by applying two stimuli at two appropriately chosen points on a ring (1 and 2, as it is shown in Fig. 6) with a time delay.

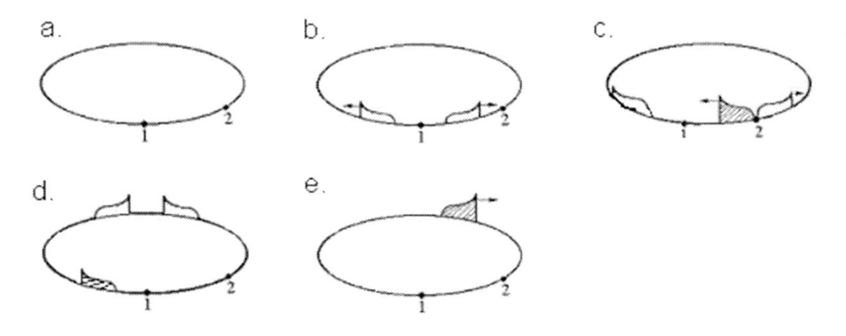

Fig. 6. Initiation of a pulse propagation in a ring-shaped fiber.

7.5.2. Ring model formation and changing it length in course of computer simulation

In computer simulation, different approaches are used for initiation the pulse circulation in a ring. At the beginning, these approaches provide the solution of equations (16) or (17) of the pulse propagation in an open-end fiber model of finite length L with boundary condition $(\frac{dV_m}{dx})_{x=0,L} = 0$. The stimulus is applied to one of the ends of a fiber and causes the propagated pulse. After this pulse propagates away from the site of initiation, the fiber ends are joined, i.e. the boundary conditions introduced earlier are replaced by the conditions of periodicity: $V_{m,x=0} = V_{m,x=L}$ and $\left.\frac{\partial V_m}{\partial x}\right|_{x=0} = \left.\frac{\partial V_m}{\partial x}\right|_{x=L}$.

It is interesting to consider the following cases:

- L large compared to the propagated wavelength and the effect of recovery process can be neglected.

- L smaller and it is necessary to consider recovery processes.

The major problems, which arise in studying the pulse propagation in a ring, are:

- Find the possible regimes of propagation
- Determine the conditions of stability of propagation
- Determine the conditions of termination of pulse circulation
- Reveal the effects of a ring length, the cell and tissue properties on the character of pulse propagation

There are three known approaches: graphical, analytical (requires some computer simulation), and pure computer simulation.

7.5.3. Graphical approach

This approach is based on the following assumptions:

- APD restitution curves measured on an isolated cell and on a cell in a ring are the same.
- The velocity of pulse propagation in a ring is constant for the duration of one turn of circulation.

The circulation of excitation in a ring is governed by two relationships. The first is called the conservation equation:

$$T_{ck} = APD_k(DI_{k-1}) + DI_k \quad ; \quad T_{ck} = \frac{L}{\theta_k} \tag{19}$$

The second relationship is called the dispersion equation:

$$\theta_k = f(DI_{k-1}) \tag{20}$$

Here:

- k subscript indicates the number of the pulse turn in a ring
- T_{ck} is a time required for circulation of the pulse around the ring
- APD_k is the action potential duration
- DI_k is the diastolic interval
- θ_k is the velocity of pulse propagation

The equation (19) represents a straight line in coordinates (APD, DI) (see fig 7b). The distance of this line from the center of coordinates is proportional to the ring length, *L*. In fig. 7b, the dependence (20) is plotted under the APD restitution curve. Both curves were obtained by computer simulation of the simplified model [14]. The first turn of pulse propagation in a ring is clarified in fig. 7A and the other three in the following text and fig.7b. Three cases are presented in fig. 7b; with ring length $L_1>L_2>L_3$, which correspond to stable circulation, circulation on the border of instability and unstable circulation respectively for the ring tissue formed with the cell models without developed Ca dynamics.

For explanation, let us study several turn of pulse circulations using these curves. Conservation Equation:

$$T_{ck} = APD_k\left(DI_{k-1}\right) + DI_k \,;\; T_{c_k} = \frac{L}{\theta_{over_k}}$$

First Turn

We initiate the excitation propagation in a ring when 1D fiber is at rest or, in other words, after an infinitely long previous diastolic interval.

Therefore, the $APD_1=APD_{1\max}$ and $\theta_1 = \theta_{1\max}$

$$\lambda_1 = \theta_{1\max} APD_{1\max}$$

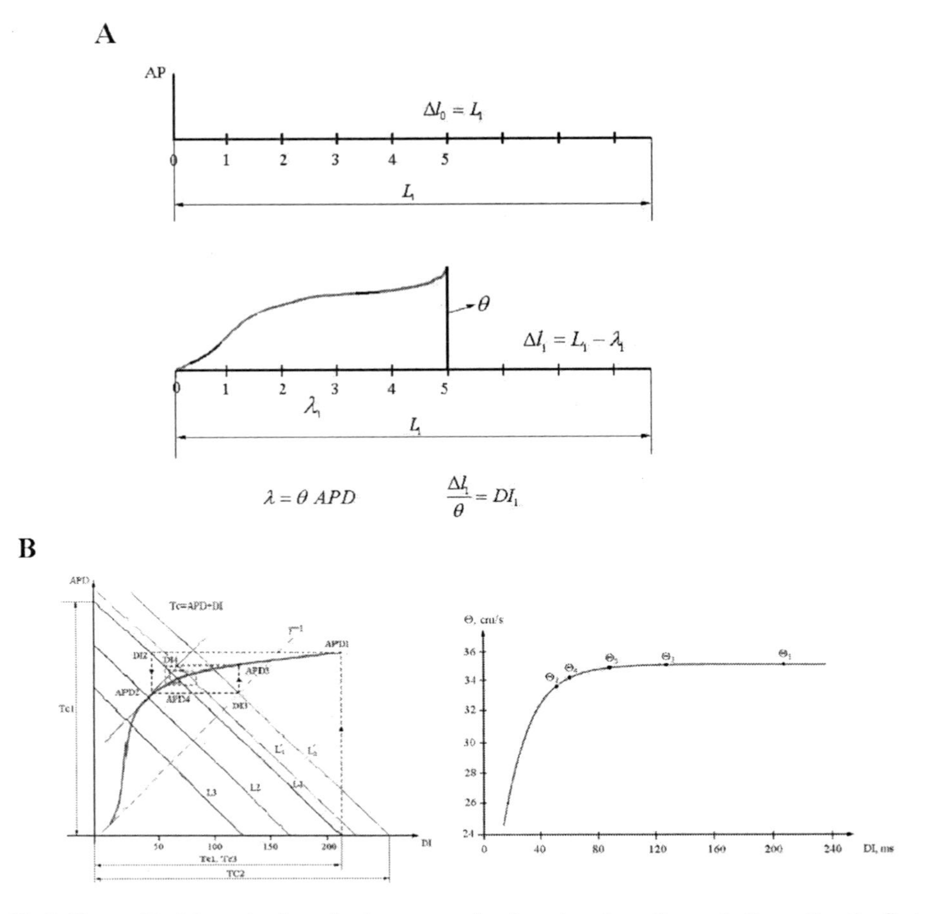

Fig.7. To graphical investigation of pulse propagation in a ring shape tissue. A. Presenting the first turn of pulse propagation. B. The three ring lengths: L_1, L_2, L_3 are shown together with APD restitution curve. L_1 corresponds to the stable circulation, L_2 to the case when circulation is on the border of stability, and L_3 when unstable circulation may lead to termination of circulation.

We assume that $L_1 > \lambda_1$, so $L_1 - \lambda_1 = \Delta l_1$ and $DI_1 = \dfrac{\Delta l_1}{\theta_{1\max}}$

Therefore, $T_{c_1} = APD_{1\max} + DI_1 = \dfrac{L_1}{\theta_{1\max}}$

Second Turn

From APD restitution and dispersion curves (see fig. 7b), we find APD_2 and θ_2. $\theta_2 < \theta_1$. Assuming that θ_2 is constant on the second pulse turn we obtain $T_{c2} > T_{c1}$ and on the graphic in Fig. 7b the straight line L changes its position to L^1_1. Using obtained values of APD_2 and θ_2, we determine the corresponding wavelength and DI_2 for the second turn:

$\lambda_2 = \theta_2 APD_2$ so $\lambda_2 < \lambda_1$

$L_1 - \lambda_2 = \Delta l_2$, so $\Delta l_2 > \Delta l_1$ and

$DI_2 = \dfrac{\Delta l_2}{\theta_2}$; $DI_2 > DI_1$

<u>Third turn</u>

Using APD restitution and dispersion curves we find:

$APD_3(DI_2)$ and $\theta_3(DI_2)$; $\theta_3 \approx \theta_1$ but $\theta_3 > \theta_2$

The third cycle time, $T_{c3} = \dfrac{L_1}{\theta_3} \approx T_{c1}$ and line $L^1{}_1$ return to the close vicinity of L_1.

The wave length $\lambda_3 = \theta_3 APD_3$. Because $APD_3 > APD_2$, $\lambda_3 > \lambda_2$ and the diastolic interval on the third turn is: $DI_3 = \dfrac{\Delta l_3}{\theta_3}$ here $\Delta l_3 = L_1 - \lambda_3$

Because $\Delta l_3 < \Delta l_2$, , $DI_3 < DI_2$ and so forth...

The graphical approach shows that it is possible to make some qualitative judgments about the circulation based on the slope of the APD restitution curve and the length of a ring,. Normally, if the APD restitution curve has slope smaller than one for all DI's, the circulation is stationary except for very short ring length ($L \le \lambda$), when it terminates.

When APD restitution curve has some interval of DI for which the slope $\gamma > 1$, it is possible to obtain stationary circulation for long ring length, unstable for the ring length which provide circulation with DI's corresponding to the slope of APD restitution curve with $\gamma \ge 1$, and termination for the shorter ring-lengths.

7.6. Analytical approach

Pure analytical solution of the whole problem is not available. Courtemanche et.al. developed an original and elegant analytical approach to determine the conditions for instability of pulse propagation in a ring [9, 10]. They also determined the transition from steady state circulation to quasi-periodic oscillatory regime and derived the expressions to estimate the parameters. The approach is based on reduction of the original PDE to the time delay neutral differential equations.

This theory is based on the assumption that both APD and velocity of pulse propagation at each point on a ring can be expressed as functions of DI in this point on the previous turn (APD restitution and dispersion curves are exist and unique).

The time conservation relations are used just like in the graphical approach, but no assumption is made about the constancy of the velocity of propagation on each turn of pulse circulation. Indeed, for each node on a ring with coordinate x, the conservation relation gives:

$$T_c(x) = APD(DI(x-L)) + DI(x) \tag{21}$$

On the other hand, the time of one pulse turn with variable velocity of propagation can be determined as:

$$T_c(x) = \int_{x-L}^{x} \frac{ds}{\theta[DI(s)]} \tag{22}$$

From (21) and (22):

$$DI(x) = \int_{x-L}^{x} \frac{ds}{\theta[DI(s)]} - APD(DI(x-L)) \tag{23}$$

Let us differentiate the both sides of (23) with respect to x:

$$\frac{d}{dx}\left(DI(x) + APD(DI(x-L))\right) = \frac{1}{\theta(DI(x))} - \frac{1}{\theta(DI(x-L))} \tag{24}$$

The equation (24) is a non-linear neutral delay equation and its solution can be found numerically if the non-linear functions $APD(DI(x-L))$, $\theta(DI(x-L))$ and the initial distribution of $DI(x)$ are known. These data can be obtained only by first solving the original PDEs. From the point of view of numerical solution, this transformation of the problem formulation makes no sense. The main goal of [15,16] was to find the condition of instability of the possible steady state solution of (24). In steady state, the velocity of propagation along a ring is constant. Designating the steady state values of $\theta=\theta^*$ and $DI=DI^*$ in (21) and (22) we obtain:

$$DI^*(x) = \frac{L}{\theta^*(DI^*)} - APD(DI^*) \tag{25}$$

$$T_c(x) = \frac{L}{\theta^*(DI^*)} \tag{26}$$

To find the stability of the possible steady state solution of (24) it is necessary to investigate its behavior close to the steady state solution. Designating the increments $y(x) = DI(x) - DI^*$, $y_L = DI(x-L) - DI^*$ and linearizing (24), we obtain:

$$\frac{dy}{dx} + \gamma\frac{dy_L}{dx} = -\alpha(y - y_L) \tag{27}$$

Here:

$$\gamma = \left(\frac{\partial APD}{\partial DI}\right)_{DI^*}$$

$$\alpha = \frac{\left(\frac{\partial \theta}{\partial DI}\right)_{DI^*}}{\theta^2(DI^*)}$$

Let us analyze the stability conditions for linear equation (27). These conditions will be true for original nonlinear equation (24) in the vicinity of the possible steady state solution. Because (27) is a linear delay equation, let us apply the Laplace transformation to both sites:

$$p + \alpha = (\alpha - \gamma p)e^{-pL} \tag{28}$$

Solving (28) with respect to the delay operator, we obtain:

$$e^{pL} = \frac{\alpha - \gamma p}{p + \alpha}$$

Here the complex variable $p = \delta + j\omega$ and therefore

$$e^{\delta L} e^{j\omega L} = \left(\frac{(\alpha - \gamma\delta) - j\omega\gamma}{(\delta + \alpha)^2 + \omega^2} \right)$$

This equality will be true if and only if:

$$e^{\delta L} = \left(\frac{(\alpha - \gamma\delta)^2 + \omega^2\gamma^2}{(\delta + \alpha)^2 + \omega^2} \right)^{\frac{1}{2}} \tag{29}$$

and

$$\omega L = -\arctan\left(\frac{\omega\gamma}{\alpha - \gamma\delta} \right) - \arctan\left(\frac{\omega}{\delta + \alpha} \right) \tag{30}$$

For linear systems, the border of stability solution is achieved when the real part of the root of the characteristic equation becomes equal to zero. In our case, this condition corresponds to $\delta = 0$. So, letting $e^{\delta L} = 1$ in (29) and

$$\left(\frac{\alpha^2 + \omega^2\gamma^2}{\alpha^2 + \omega^2} \right)^{\frac{1}{2}} = 1$$

The equation holds when $\gamma = 1$. It means that the condition of instability is $\gamma \geq 1$. For $\gamma = 1$ the phase condition (30) transformed to:

$$\omega L = -2\arctan\frac{\omega}{\alpha}$$

Assuming $\alpha L << 1$, in [15] this implicit equation was solved approximately and all ω_k and corresponding periods $\Lambda_k = \frac{2\pi}{\omega_k}$ of space distribution of *DI(x)* were obtained from the expression:

$$\Lambda_k = \frac{2L}{2k+1} - \frac{4L^2\alpha}{(2k+1)^3\pi^2} \qquad k = 0,1,2,3.... \tag{31}$$

Among all solutions of Λ_k found to date, only two have been observed, namely for *k=0* and *k=1*. The second term in (31) is comparatively small and determines the appearance of the beats in DI and APD for any node of a ring. The periods of the beats are equal 2L/2L- Λ_k. This form of oscillation is called quasi-periodic.

According to the Hopf bifurcation theory, the appearance of an oscillating solution in a linearized system when $\gamma = 1$ guarantees the appearance of an oscillating solution in the original nonlinear system when the DI in close to the DI^*. This theorem provides no proof of stability of this oscillatory solution for original nonlinear system for DI significantly different from the DI* and cannot be used to describe the system behavior for shorter ring-length.

7.7. APD and velocity restitution in a ring-shaped model

The graphical and analytical methods described in previous paragraphs are based on assumptions that APD restitution curve and velocity of wave propagation exist at any point of the tissue and are the single-valued functions of previous DI.

The restitution properties express an ability of a cardiac cell to recover after excitation. The recovery processes are difficult if not impossible to observe during physiological experiments since they are determined by the temporal activity of membrane channels. That explains why physiologists prefer to measure the secondary effects of these processes on the duration of AP. The protocol of these measurements specifies that tissue (or its mathematical model) is preconditioned by applying periodic stimulation with a period equal to the normal heart rate until steady-state is attained. Then, after a relatively short DI, a premature excitation is applied, and the resulting APD is measured. This process is repeated from the beginning for longer initial DI.

For a long time that the dependence of APD on previous DI (APD restitution curve) was thought to be single valued. However, it was found [17] that the APD restitution curve changes when the frequency of the precondition stimulation is increased. Moreover, it was shown that different measurement protocols (e.g.,S1, S2, S3 protocol) lead to the appearance of families of APD restitution curves.

Thus, it is possible to conclude that the APD restitution curve is not a function only of the previous DI, but of the history of the preceding sequence of excitations.

Justification of these principles may be found in [18], where they were verified on the cell mathematical models based on clamp-experiment data, which reflect the dynamics of membrane channels during and after excitation. In order to check whether the APD after an excitation depends only on the previous DI, a sequence of three excitation stimuli was applied to the different single-cell models . In these simulations, the last two premature stimuli appear after equal DIs.

The results, presented in Table 1, show that only for the Nobel model after equal DIs the APs appear with equal durations. In the BR model, the APD after the second premature beat is longer than the first, whereas for the LR I in the same situation, the APD is shorter. In addition, the APD restitution curves are measured for these models, using computer simulations performed using S1, S2, and S3 protocol.

Table 1. Results of consecutive stimulation with equal DIs

	Original AP		AP after first premature stimulus		AP after second premature stimulus	
Model	DI_0 (ms)	APD_0 (ms)	DI_0 (ms)	APD_0 (ms)	DI_0 (ms)	APD_0 (ms)
Nobel	∞	380	38	151	38	151
BR	∞	291	20	117	20	146
LR I	∞	384	20	256	20	242

Under these pacing conditions, instead of one APD restitution curve we have a family of curves. Remarks that APD is not a function of only the previous DI are also found in [17]. The pacing order of a cell placed in a ring will be different in comparison to these two protocols, especially during the propagation with quasiperiodic oscillation. Indeed there is experimental evidence [6] supporting that idea. In [6] during unstable circulation, different values of APD were observed for the same values of DI. Therefore, one can expect the APD restitution curve of a cell in a ring to be significantly different from that of an isolated cell. Computer simulation results reported in [18] show that the APD restitution curve of a cell in ring-shaped tissue is below that obtained for isolated cell demonstrating the effect of local current.

In summary it is necessary to underline that presented stability conditions for wave propagation in ring-shaped tissue are evaluated for the mathematical models without or with very primitive Ca dynamics (see [17, 19] in chapter 4).
Moreover, they are obtained assuming that cell's memory effect (not single valued APD restitution curve) can be neglected, and for linearized originally nonlinear system under condition of small perturbation in DI. Thus, these stability conditions can be considered necessary but not sufficient for the original dynamic system. It means that locally our system may be stable but globally unstable. The developed Ca dynamics significantly affect the wave propagation in a ring-shaped tissue involving new phenomena described in the next paragraph.

7.8. Propagation instability in ring-shaped tissue with Ca^{2+} dynamics

As mentioned in chapter 4, Ca^{2+} dynamics play a major role in the excitation-contraction coupling. Ca^{2+} dynamics also significantly affects the characteristics of the generated AP. The latter are caused by the effects of Na-Ca exchanger, L-type Ca channel currents, and Ca release from the SR, which activates others Ca-dependant membrane currents. These effects are increased significantly during stimulation with high pacing rates, when Ca accumulation in SR and myoplasm creates conditions for spontaneous release of Ca from SR. These conditions appear only when an excitation wave circulates in a ring-shaped tissue of sufficiently short length for some finite time.

In 1D ring of tissue, the effective period of reentrant wave circulated with a constant velocity is directly proportional to the length of the ring. Stationary propagation is observed in a relatively long ring (L>20 cm). Unstable irregular propagation, characterized by variability in the local period of excitation, was observed [19] in intermediate ring lengths (L<15–20 cm) (APD and velocity of propagation see figs 8 and 9 respectively).

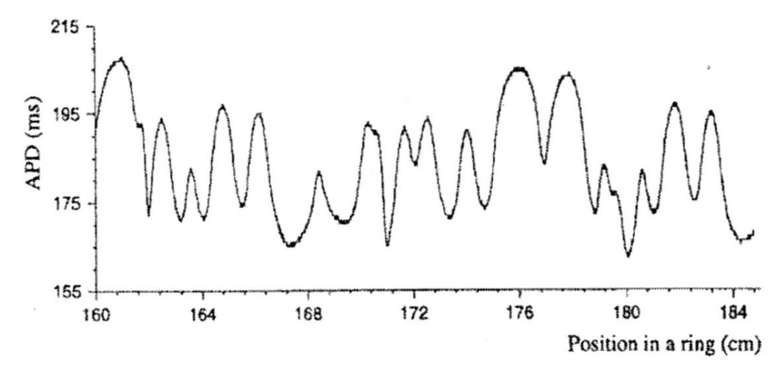

Fig. 8. APD distribution along a ring-shaped 1D cable of myocardium.

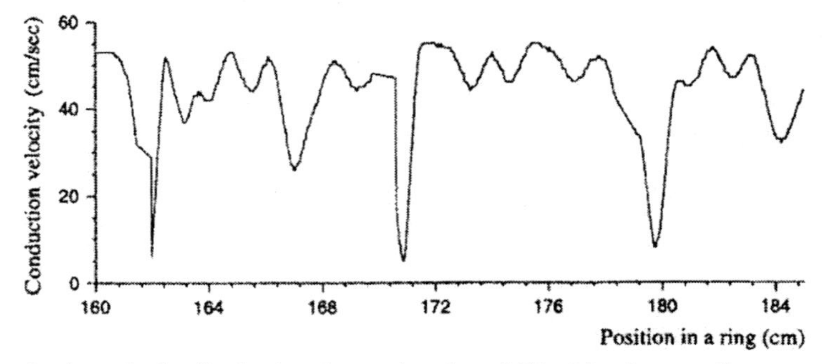

Fig. 9. Conduction velocity distribution along a ring-shaped 1D cable of myocardium.

The APD restitution was measured during the irregular regime and remained a random collection of points (fig. 10).

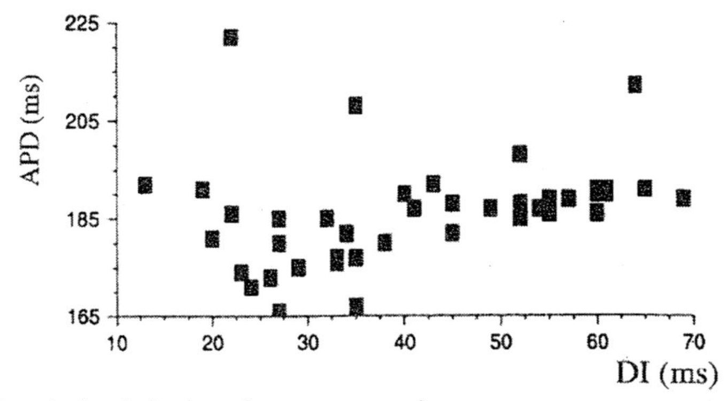

Fig. 10. APD restitution during irregular wave propagation.

At short ring lengths (L <9.728–15 cm), the rate of excitation was rapid enough to intermittently cause non-uniform Ca accumulation in the cells along a ring and following spontaneous late-diastolic SR Ca^{2+} release in cells with sufficiently large Ca accumulation. This spontaneous Ca release facilitates the appearance of an EAD on the ensuing AP. The inset in Fig. 11, demonstrates [20] that the EAD did not coincide temporally with the J_{spon} peak.

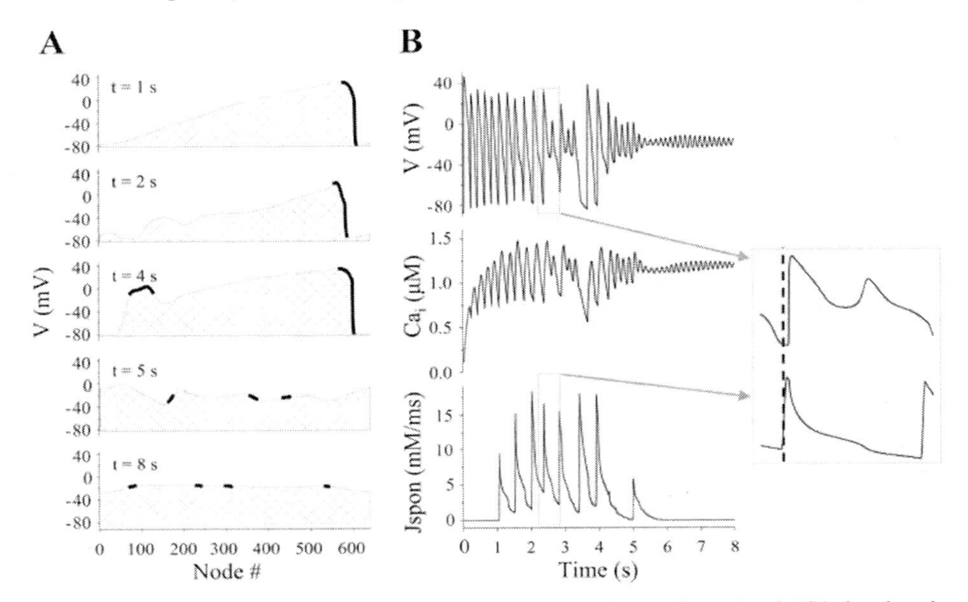

Fig.11. Spatiotemporal profiles of V_m during wave propagation in a 1-dimensional (1D) ring, length 640 nodes (10.24 cm) ***A***: spatial distribution of V_m across the ring at different times as indicated. Thick lines indicate nodes in the ring with $dV_m/dx > 0$ at the chosen time. Stationary wave propagation [time (t) = 1 s] becomes irregular (t = 2 s), and then early afterdepolarizations (EADs) appear (t = 4 s) as bumps on the tail of the wave with $dV_m/dt > 0$. After 5 s, the tissue enters a state of repolarization failure and never returns to rest potential (t = 8 s). ***B***: traces of V_m, intracellular Ca (Ca$_i$), and spontaneous Ca release (J_{spon}) from SR are taken at node 160 during wave circulation. As time passes, V_m trace show EAD activity becoming more pronounced, from single EADs to multiple EADs and to the state of repolarization failure. Dashed vertical line in the inset indicates that spontaneous SR Ca release occurs during late diastole preceding the action potential (AP) in which an EAD occurs.

Spontaneous late-diastolic SR Ca release triggered the EAD during the ensuing AP, consistent with experimental observations of DADs preceding the upstroke of APs exhibiting EADs [21, 22]. The spontaneous late diastolic Ca release combined with the $I_{Ca,L}$-induced SR Ca release produced by the ensuing AP, augmenting the Ca transient amplitude. The resultant larger Ca transient enhanced Ca-sensitive inward currents, specifically the Na/Ca exchange current (I_{NaCa}) and $I_{ns(Ca)}$, during the AP plateau phase. The enhanced inward currents decreased repolarization reserve and thereby established a more tenuous balance of repolarizing currents, such that window $I_{Ca,L}$ reactivation was able to generate an EAD. If J_{spon} in the cell model was inactivated, EADs did not develop. Although every EAD was preceded by a corresponding late diastolic J_{spon} peak, not every J_{spon} peak was followed by an EAD (fig. 11*B*). For very short ring lengths (L= 9.728 cm), the ring was not sufficiently long to sustain reentry. Here, we consider only those ring lengths for which Ca release-induced EADs occurred (L=9.728 –15 cm).

Fig. 11 A shows the spatial distribution of membrane voltage along the ring at different moments in time, while Fig 11B shows the voltage, Ca_i, and J_{spon} traces of an arbitrary node in the ring. After a short transient period during the first few turns of the wave, stationary propagation is established ($t \approx 1$ s; Fig. 11B). From 1 to 2 s, spontaneous diastolic SR Ca releases (because of activations of J_{spon}) began and delayed repolarization appears, indicating diminished repolarization reserve. However, at this point, the spontaneous SR Ca release events were not sufficiently large to produce EADs ($t \approx 2$ s; Fig. 11B). As it was shown [19, 23], the repolarization delay was heterogeneous because of an inhomogeneous spatial distribution of SR Ca accumulation, and hence J_{spon} amplitude, along the ring. From 2–5 s, spontaneous diastolic SR Ca release resulting from J_{spon} activation became large enough to induce EADs in various regions of the ring. These EADs could both terminate and regenerate wave propagation. EADs are seen in voltage traces from single nodes (fig. 12*B*) and appear as bumps with positive dV_m/dx on the tail of the voltage wave, indicated in the spatial distribution of voltage (Fig. 2*A*) by the thick lines. EADs appeared slightly earlier in shorter rings. Multiple EADs are also observed during this time (fig. 12*B*).

EADs affected wave propagation in several ways. A region of EADs could stop a propagating wave if the region was sufficiently large and arose just ahead of the wavefront, or it could regenerate wave propagation if it was sufficiently large and arose adjacent to a region of repolarized tissue into which the new wave could propagate. Alternatively, a region of EADs could occur in such a way that propagation was terminated and not regenerated. For example, a region of EADs could arise that prolonged refractoriness enough to block reentry of the original wave but was not large enough to trigger a new wave. Wave regeneration could be prevented manually by inactivating J_{spon} in all cell models just after the original wave stopped. Several different regions of EADs could also arise in the ring at the same time ($t \approx 5$ s; fig. 12*A*).

Simulation experiments in [23] described EAD-induced regeneration of wave propagation antegradely. The existence of the following new modes of EAD-induced wave regeneration was demonstrated [20]:

- a new wave traveling retrogradely;
- two new waves traveling in the same direction, both antegrade (Fig. 12*A*);
- two new waves traveling in opposite directions, antegrade and retrograde(Fig. 12*B*),
- a special case of the last mode, where two waves traveling in opposite directions arose from a single region of EADs, which then propagated in both directions.

In the time window from 2 to 5 s and within the range of ring lengths where EADs occurred, we observed any or all of these four modes of wave regeneration depending upon the ring length.

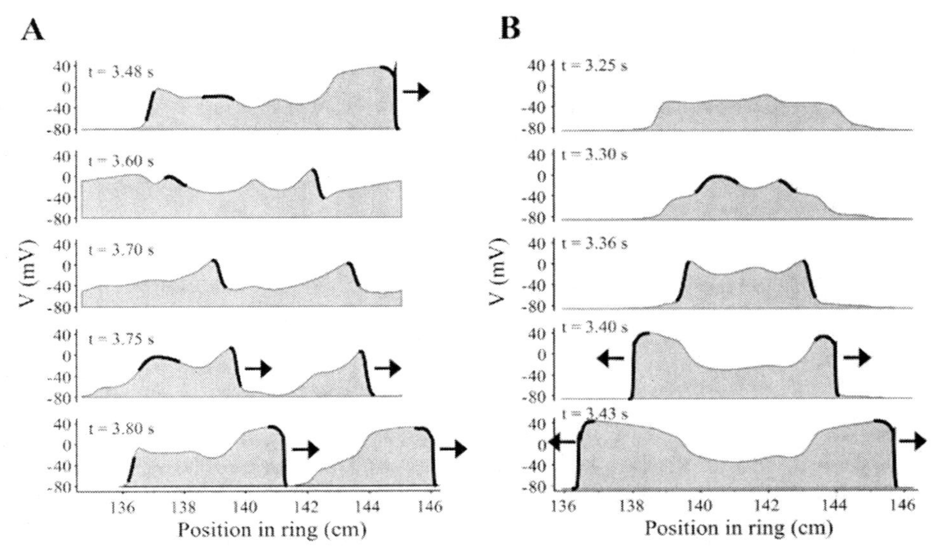

Fig . 12. EAD-induced regeneration of two simultaneous waves in a 1D ring. The spatial profile of V_m is shown at different times, with the thick lines indicating regions with $dV_m/dt > 0$. The original circulating wave front is indicated by the arrow in panel on *top*, and other regions with $dV_m/dt > 0$ correspond to EADs (no arrow) or EAD-induced propagating wave fronts (arrows in subsequent panels). ***A*:** regeneration of two waves in the same direction in a ring of 650 nodes (10.4 cm). After termination of the initial wave by EADs, two distinct regions of EADs occur, each bordering repolarizing tissue (*t* _3.70 s). Two antegrade wave fronts generate (*t* _ 3.75 s), although they eventually collide and terminate each other since the ring is not long enough to sustain two fully propagated waves. ***B***: similar to *A*, except that the regenerated waves propagate in opposite directions and extinguish by collision of wave fronts. Ring length was 645 nodes (10.32 cm).

Changing the ring length altered the variability in the local period of excitation during reentry and led to different patterns of regional pacing history. The subsequent patterns of inhomogeneous J_{spon} distribution along the ring determined which modes of wave regeneration occurred. The chaotic nature of the reentry makes the prediction of the regeneration wave difficult if not impossible. Unfortunately, the author did not find a treatment of the effect of $[Ca^{2+}]_i$ dynamics on wave propagation along ring-shaped cardiac tissue in current literature.

7.9. References

1. Karma, A., H. Levine, and X. Zou, Theory of pulse instability in electrophysiological models of excitable tissues. Physica D, 1994. **73**: 113-127.
2. Chialvo, D.R., R.F.J. Gilmour, and J. Jalife, Low dimensional chaos in cardiac tissue. Nature, 1990. **343**: 653-657.
3. Ito, H. and L. Glass, Theory of reentrant excitation in a ring of cardiac tissue. Physica D, 1992. **56**: 84-106.
4. Frame, L.H. and M.B. Simson, Oscillations of conduction, action potential duration, and refractoriness. A mechanism for spontaneous termination of reentrant tachycardias. Circulation, 1988. **78**: 1277-1287.

5. Fei, H., M.S. Hanna, and L.H. Frame, Assessing the excitable gap in reentry by resetting: implications for tachycardia termination by premature stimuli and antiarrhythmic drugs. Circulation, 1996. **94**: 2268-2277.
6. Fei, H., D. Yazmajian, M.S. Hanna, and L.H. Frame, Termination of reentry by lidocaine in the tricuspid ring in vitro: role of cycle-length oscillation, fast use-dependent kinetics, and fixed block. Circ Res, 1997. **80**: 242-252.
7. Rudy, Y. and W.L. Quan, A model study of the effects of the discrete cellular structure on electrical propagation in cardiac tissue. Circ Res, 1987. **61**: 815-23.
8. Keener, J. and J. Sneyd, Mathematical Physiology. 2nd ed. 2001: Springer-Verlag.
9. Hodgkin, A.L. and A.F. Huxley, A quantitative description of membrane current and its application to conduction and excitation in nerve. J Physiol, 1952. **117**: 500-544.
10. Sakmann, B. and E. Neher, eds. Single Channel Recording. 1983, Plenum Press: New York.
11. Panfilov, A.V. and A.V. Holden, eds. Computational Biology of the Heart. 1997, Wiley Publishing: New York.
12. Nagumo, J., S. Arimoto, and S. Yoshizawa, An active pulse transmission line simulating nerve axon. Proceedings of the IRE, 1962. **50**: 2061-2070.
13. Zykov, V.S., Simulation of Wave Process in Excitable Media. Nonlinear science: theory and applications, ed. A.V. Holden. 1987, Manchester and New York: Manchester University Press.
14. Kogan, B.Y., W.J. Karplus, and M.G. Karpoukhin, The third-order action potential model for computer simulation of electrical wave propagation in cardiac tissue., in Computer Simulations in Biomedicine, H. Power and R.T. Hart, Editors. 1995, Computational Mechanics Publishers: Boston.
15. Courtemanche, M., L. Glass, and J.P. Keener, Instabilities of a propagating pulse in a ring of excitable media. Phys Rev Lett, 1993. **70**: 2182-2185.
16. Courtemanche, M., J.P. Keener, and L. Glass, A delay equation representation of pulse circulation on a ring of excitable media. SIAM J Appl Math, 1996. **56**: 119-142.
17. Franz, M.R., J. Schaefer, M. Schottler, W.A. Seed, and M.I.M. Noble, Electrical and mechanical restitution of the human heart at different rates of stimulation. Circ Res, 1983. **53**: 815-822.
18. Kogan, B.Y., W.J. Karplus, M.G. Karpoukhin, I.M. Roizen, E. Chudin, and Z. Qu, Action potential duration restitution and electrical excitation propagation in a ring of cardiac cells. Comput Biomed Res, 1997. **30**: 349-359.
19. Chudin, E. and B. Kogan, Pulse propagation in a ring-shaped cardiac tissue model with intracellular Ca(2+) dynamics. (Computer simulation study), in Mathematics and Computers in Modern Science: Acoustics and Music, Biology and Chemistry, Business and Economics, N. Mastorakis, Editor. 2000, World Scientific and Engineering Society Press: Athens, Greece. p. 187-192.
20. Huffaker, R.B., J.N. Weiss, and B. Kogan, Effects of early afterdepolarizations on reentry in cardiac tissue: a simulation study. Am J Physiol Heart Circ Physiol, 2007. **292**: H3089-H3102.
21. Priori, S.G. and P.B. Corr, *Mechanisms underlying early and delayed afterdepolarizations induced by catecholamines.* Am J Physiol, 1990. **258**: H1796-H1805.
22. Volders, P.G., A. Kulcsar, M.A. Vos, K.R. Sipido, H.J. Wellens, R. Lazzara, and B. Szabo, *Similarities between early and delayed afterdepolarizations induced by isoproterenol in canine ventricular myocytes.* Cardiovasc Res, 1997. **34**: 348-359.
23. Huffaker, R., S.T. Lamp, J.N. Weiss, and B. Kogan, *Intracellular calcium cycling, early afterdepolarizations, and reentry in simulated long QT syndrome.* Heart Rhythm, 2004. **1**: 441-448.

Chapter 8. Waves in Two Dimensional Models of Myocardium

Normal heart function is directly connected with periodic propagation of excitation waves initiated by the pacemaker heart system. Electrophysiological experiments show that distortions in heart rhythm such as tachycardia are a precursor to ventricular fibrillation (see Fig. 5A and 5B in Chapter 1, Introduction). Ventricular fibrillation may occur in either already damaged or initially healthy hearts. The mechanisms of ventricular fibrillation are not fully understood. In current literature [1], monomorphic tachycardia is associated with stationary propagation of spiral excitation waves, while polymorphic tachycardia is thought to be due to non-stationary propagation. The breakup of a wavefront of a non-stationary propagating spiral wave obtained in computer simulation with tissue formed of AP models without Ca dynamics is considered fibrillation [2]. Spiral waves were discovered during computer simulations [3]. Their existence was confirmed, years later, in the course of physiological experiments [4] in 2D normal atrium cardiac tissue by properly applied premature stimulation.

Generally, it is convenient to divide excitation-propagation into two cases: wave propagation with rectilinear and curvilinear fronts. Waves with rectilinear fronts represent a particular case of curvilinear, when the radius of curvature along wavefront tends to infinity equally for all the front points. Spiral waves represent one of the most important types of traveling waves. Unfortunately, theory has only been developed for stationary spiral waves ([5], [6]). Here and in the next chapters, we discuss the elements of stationary spiral wave theory, preliminarily introducing some important assumptions about cardiac tissues (e.g. representation as a continuous media – a syncytium) and some consideration about mono- and bidomain approaches used in mathematical modeling of cardiac tissues.

8.1. Heart muscle as a 2D and 3D syncytium

The idea of a syncytium as continuous mono- and bidomain media can be illustrated by considering the limit to which a uniform and discrete 2D grid (fig. 1) tends to when the distance between the grid nodes goes to zero.

B.Ja. Kogan, *Introduction to Computational Cardiology: Mathematical Modeling and Computer Simulation*, DOI 10.1007/978-0-387-76686-7_8,

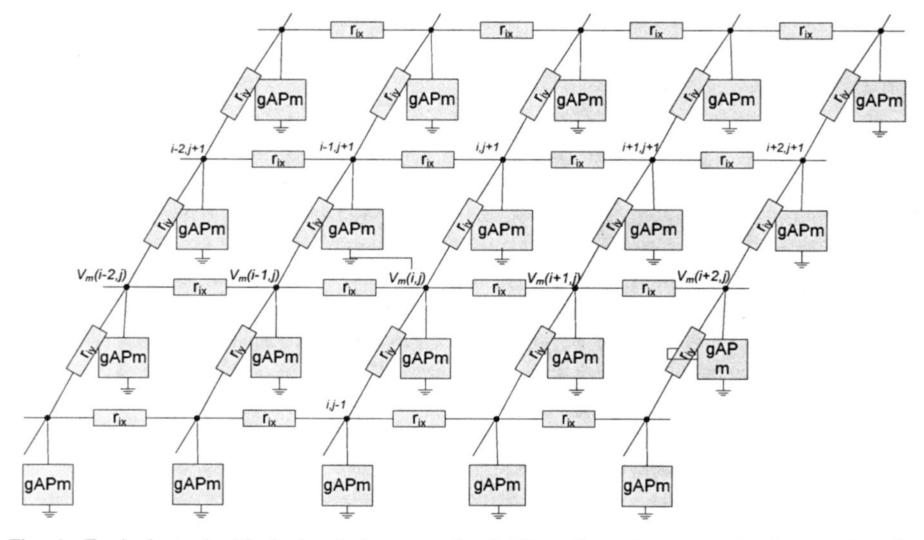

Fig. 1. Equivalent electrical circuit for a patch of 2D uniform heart muscle tissue using the monodomain approach (the resistance of extracellular liquid is much smaller than intracellular). gAPm designates a generalized AP model. The intracellular coupling resistances in the x and y directions are respectively r_{ix} and r_{iy}.

According to the electrical equivalent circuit diagram of a tissue model shown in fig. 1, the external local current for the node located at (i, j) is:

$$i^*_{ext_{i,j}} = \frac{V_{i+i,j} - V_{ij}}{r_{ix}} + \frac{V_{(i-1)j} - V_{ij}}{r_{ix}} + \frac{V_{i,j+1} - V_{i,j}}{r_{iy}} + \frac{V_{i(j-1)} - V_{ij}}{r_{iy}}$$

Here:

$V_{i,j}$ – is a membrane voltage V_m at the node located at (i, j),

r_{ix} and r_{iy} – are the coupling resistances along the longitudinal and transverse cell axes.

Reordering the previous equation, we obtain:

$$i^*_{ext_{i,j}} = \frac{V_{(i+1)j} + V_{(i-1)j} - 2V_{ij}}{r_{ix}} + \frac{V_{(j+1)i} + V_{(j-1)i} - 2V_{ij}}{r_{iy}} \tag{1}$$

The dimensions of r_{ix} and r_{iy} is [kΩ-cm^2].

Let us introduce the following substitutions in (1):

$r_{ix} = K_x \Delta x^2$ and $r_{iy} = K_y \Delta y^2$ (dimension of K_x and K_y is kΩ).

This gives:

$$i^*_{ext_{ij}} = \frac{V_{(i+1)j} + V_{(i-1)j} - 2V_{ij}}{K_x \Delta x^2} + \frac{V_{(j+1)i} + V_{(j-1)i} - 2V_{ij}}{K_y \Delta y^2}$$

Taking the limit of both sides of this equation, as Δx and Δy both tend to zero:

$$\lim_{\Delta x, \Delta y \to 0} i^*_{ext_{ij}} = \frac{1}{K_x} \lim_{\Delta x \to 0} \frac{V_{i+1,j} + V_{(i-1)j} - 2V_{ij}}{\Delta x^2} + \frac{1}{K_y} \lim_{\Delta y \to 0} \frac{V_{(j+1)i} + V_{(j-1)i} - 2V_{ij}}{\Delta y^2}$$

We obtain i_{ext} expressed as a Laplacian of V_m:

$$i_{ext} = \frac{1}{K_x}\frac{\partial^2 V_m}{\partial x^2} + \frac{1}{K_y}\frac{\partial^2 V_m}{\partial y^2} \tag{2}$$

According to fig. 2, this external current must be equal to total membrane ionic and capacitance currents:

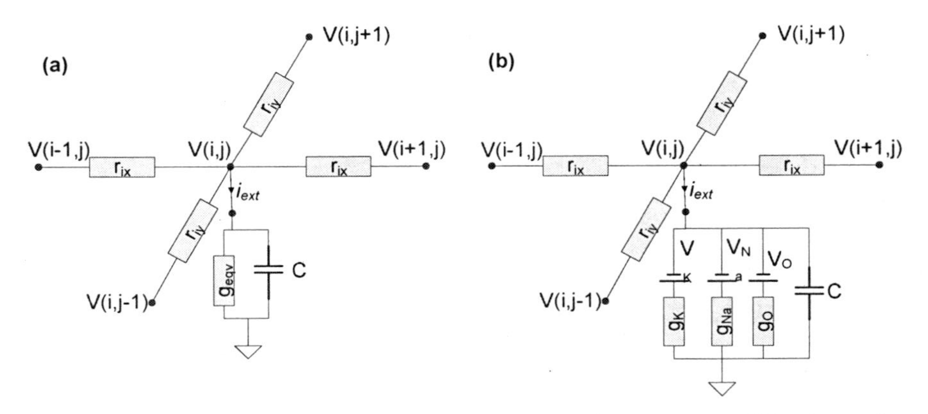

Fig. 2. Generalized AP equivalent circuit diagrams in 2D tissue for: a) passive and b) active modes.

$$i_{ext} = C_m \frac{\partial V_m}{\partial t} + \sum I_S - I_{St} = \frac{1}{K_x}\frac{\partial^2 V_m}{\partial x^2} + \frac{1}{K_y}\frac{\partial^2 V_m}{\partial y^2}$$

After division by C_m and denoting:

$\frac{1}{C_m K_x} = D_x\left[\frac{cm^2}{\sec}\right]; \quad \frac{1}{C_m K_y} = D_y\left[\frac{cm^2}{\sec}\right]$, the mathematical model for 2-D monodomain excitable tissue becomes:

$$D_x \frac{\partial^2 V_m}{\partial x^2} + D_y \frac{\partial^2 V_m}{\partial y^2} = \frac{\partial V_m}{\partial t} + \frac{\sum I_S + I_{St}}{C_m} \tag{3}$$

It is evident that for three-dimensional uniform tissues (3D) in the monodomain representation, the expression (3) is changed only by the Laplacian:

$$D_x \frac{\partial^2 V_m}{\partial x^2} + D_y \frac{\partial^2 V_m}{\partial y^2} + D_z \frac{\partial^2 V_m}{\partial z^2} = \frac{\partial V_m}{\partial t} + \frac{\sum I_S + I_{St}}{C_m} \tag{4}$$

Here, we assume that the diffusion coefficients D_x, D_y, and D_z are different but constant along their coordinate axes.

If $I_{St} > I_{th}$, then expressions (3) and (4) describe active propagation. If $I_{St} < I_{th}$ and $\sum I_s$ is replaced by $g_{eqv} V_m$ in these equations, then they describe passive propagation.

Both equations (3) and (4) must be solved with appropriate boundary and initial conditions. Neumann boundary conditions are typically used. Neuman boundary conditions are characterized by no current flow across the boundaries (i.e.

$\left(\frac{\partial V_m}{\partial n}\right)_B = 0$). Initial conditions are[1]: $V_m(0)=V_{rest}$ and all $y_i(0)$ and $[S](0)$ reflect the properties of the utilized AP model. For passive propagation, boundary conditions remain the same, but the initial condition is reduced to $V_m(0) = V_{rest}$.

8.1.1. Anisotropy of the tissue

Isotropic 2D tissue is characterized by equal diffusion properties in both geometrical directions, $D_x = D_y = D$, with the Laplacian (L) in (3) reducing to:

$$L = D\left(\frac{\partial^2 V_m}{\partial x^2} + \frac{\partial^2 V_m}{\partial y^2}\right) \tag{5}$$

Anisotropic tissue has, generally, $D_x \neq D_y$. For heart muscle, $D_x > D_y$; namely $D_x \approx$ (3 or 4)× D_y. We distinguish two cases: the first is uniform anisotropy, when D_x and D_y do not change with the space coordinates x and y respectively; and the second is non-uniform anisotropy, when the diffusion coefficients are a function of these coordinates ($D_x(x) \neq D_y(y)$). In the first case, it is possible to transform the Laplacian from the form used in (3) to that in (5) by scaling one of the space coordinates. Indeed:

$$\left(D_x \frac{\partial^2 V_m}{\partial x^2} + D_y \frac{\partial^2 V_m}{\partial y^2}\right) = D_x\left(\frac{\partial^2 V_m}{\partial x^2} + \frac{D_y}{D_x}\frac{\partial^2 V_m}{\partial y^2}\right)$$

and designating: $\frac{D_y}{D_x} = d$, $\hat{y} = \frac{y}{d}$ we reduce the expression for the Laplacian to the form (5), but with scaled space coordinate $\hat{y}$. All of the values obtained from computer simulations with this transformed Laplacian, which depend on coordinate y, must be rescaled by multiplying them by the scale factor d.

In the second case, the Laplacian in the left side of (3) will take the form:

$$\frac{\partial}{\partial x}(D_x(x)\frac{\partial V_m}{\partial x}) + \frac{\partial}{\partial y}(D_y(y)\frac{\partial V_m}{\partial y}) \tag{6}$$

In real cardiac tissue, non-uniformity also exists due to variability of the longitudinal directionality of the fibers (see [7]). This leads to representation of the diffusion coefficient as a function of conductivity tensors. The curvilinear nature of fibers can be neglected, as a first-order approximation, for small tissue pieces. In 3D tissue, transmural heterogeneity is also present (see chapter 3).

---

[1] To find the initial conditions (ICs) for phase variables in a particular AP mathematical model, it is possible to use the results of computer simulation of this model with a normal pacing rate and qualitatively selected ICs. The system must be stable and the steady state solutions of AP phase coordinates have to show all real values of unknown ICs.

8.2. Bidomain representation of 2D tissue

Mathematical models for computer simulations of cardiac tissue with a bidomain representation [8] were developed to study wave propagation under application of excitation stimuli to the external tissue domain. That is very important in cases of defibrillation processes simulations (when the electrical shock is applied to the surface of the body or to the pericardium) and when waves initiated by the pacemaker system due to some abnormal processes meet extracellular liquid with resistance comparable with that of intracellular. In all of these situations, the myocardium is considered as a syncytium [9]. The bidomain approach for studying excitation-propagation processes in tissue, in the presence of an external stimulus (particularly, a defibrillation shock), gives rise to the following generalized system of equations [10]:

$$\nabla \cdot (\hat{\sigma}_i \nabla \Phi_i) = \beta_{SV} I_m = \beta_{SV} \left(C_m \frac{\partial V_m}{\partial t} + I_{ion} + I_{i\,stim} \right) \tag{7a}$$

$$\begin{aligned} \nabla \cdot (\hat{\sigma}_e \nabla \Phi_e) &= -\beta_{SV} (I_m + I_{estim}) \\ &= -\beta_{SV} \left(C_m \frac{\partial V_m}{\partial t} + I_{ion} + I_{i\,stim} \right) - \beta_{SV} I_{estim} \end{aligned} \tag{7b}$$

$$V_m = \Phi_i - \Phi_e, \tag{7c}$$

with homogeneous Neumann boundary conditions and appropriate initial conditions.

In (7a) and (7b), $\hat{\sigma}_i$ and $\hat{\sigma}_e$ are the conductivity tensors for the intracellular and extracellular domains, respectively (mS/cm). They reflect the variable fiber directionality in 3D tissue. Other variables in (7) are:

- β_{sv} – the myocyte surface-to-volume ratio (cm^{-1});
- C_m – membrane capacitance per unit area [$\mu F/cm^2$];
- I_{ion} – the sum of transmembrane currents [$\mu A/cm^2$];
- I_{istim} and I_{estim} – stimulus currents applied to the intra- and extracellular surfaces of a membrane [$\mu A/cm^2$];
- Φ_i and Φ_e –intra- and extracellular potentials correspondingly [mV];
- V_m – difference of membrane potentials [mV];
- t – time [ms].

The system of equations (7) can be simplified by excluding the variable Φ_i from (7a) by substituting for it the value from (7c), followed by the summation of the obtained equation with (7b). As a result, we obtain the general form for the bidomain representation of cardiac tissue:

$$\frac{\partial V_m}{\partial t} = \frac{1}{\beta_{SV} C_m} (\nabla(\sigma_i \nabla V_m) + \nabla(\sigma_i \nabla \Phi_e)) - \frac{1}{C_m} (I_{ion} + I_{Istim}) \tag{8}$$

$$\nabla((\sigma_i + \sigma_e) \nabla \Phi_e = -\nabla(\sigma_i \nabla V_m) - \beta_{SV} I_{eStim} \tag{9}$$

Here ∇ is a space gradient operator. For 1D, 2D, and 3D, these operators are equal to $\nabla_1 = i\frac{\partial}{\partial x}$; $\nabla_2 = i\frac{\partial}{\partial x} + j\frac{\partial}{\partial y}$; $\nabla_3 = i\frac{\partial}{\partial x} + j\frac{\partial}{\partial y} + \kappa\frac{\partial}{\partial z}$, respectively

Assuming no fiber curvature but recognizing the presence of uniform anisotropy in both domains between the longitudinal and transverse directions, (8) and (9) for 2D tissue reduce to the following equations:

$$\frac{\partial V_m}{\partial t} = D_{i,x}\frac{\partial^2 V_m}{\partial x^2} + D_{i,y}\frac{\partial^2 V_m}{\partial y^2} + D_{i,x}\frac{\partial^2 \Phi_e}{\partial x^2} + D_{i,y}\frac{\partial^2 \Phi_e}{\partial y^2} - \frac{(I_{ion} + I_{stim})}{C_m} \quad (10)$$

$$\frac{\partial^2 \Phi_e}{\partial x^2} + \alpha\frac{\partial^2 \Phi_e}{\partial y^2} = \gamma_1\frac{\partial^2 V_m}{\partial x^2} + \gamma_2\frac{\partial^2 V_m}{\partial y^2} - \frac{\beta_{SV} I_{eStim}}{g_{i,x} + g_{e,x}} \quad (11)$$

In (10)-(11), $D_{i,x} = g_{i,x}/(\beta_{sv}C_m)$ and $D_{i,y} = g_{i,y}/(\beta_{sv}C_m)$ are the diffusivities along the x and y-axes, respectively [cm^2/ms]. $\gamma_1 = -g_{i,x}/(g_{i,x} + g_{e,x})$, $\gamma_2 = -g_{i,y}/(g_{i,x} + g_{e,x})$, $\alpha = (g_{i,y} + g_{e,y})/(g_{i,x} + g_{e,x})$, $g_{i,x}$, $g_{e,x}$, $g_{i,y}$, and $g_{e,y}$ are conductances [mS/cm]. The subscripts i, e refer to intracellular, extracellular domains, respectively, . The subscript x indicates the longitudinal direction of the fiber and subscript y the transverse direction.

A system of nonlinear ordinary differential equations that describes all membrane current components of I_{ion} and relevant intracellular compartment processes is needed to make (4) a closed form expression. For this purpose, it is possible to select, among existing AP models of different species, one that is adequate to solving the problem and is based on up-to-date physiological data. For example, I_{ion} in the guinea pig ventricular AP model proposed by Luo and Rudy [11] and its modifications introduced by Chudin [12] and supplements made by Huffaker and Samade [13] are suitable to study propagation phenomena under conditions of Ca_i overload in the myoplasm and SR.

8.3. Heart muscle as a system of parallel interconnected cables

The approach presented in the previous paragraph is based on the supposition that it is possible to simulate cardiac tissue as continuous media (mono- or bidomain). This is correct when gap junction resistance between the cells is negligible in comparison with the resistance of the cellular domains. There is physiological evidence that this is not true in some pathological cases (e.g. for tissue cells in a region of local ischemia). For these cases, it is theoretically possible to use the original approach proposed by Rudy and Quan in [14] and briefly described in chapter 7. Unfortunately, this approach is now computationally tractable only in 1D monodomain tissues of restricted length.

Leon and Roberge [15,16] proposed to use a system of parallel interconnected cables (see fig. 3) as a representation of cardiac fibers in 2D monodomain tissue. This abstraction is closer to real cardiac tissue topology. They neglected the gap junction resistance in comparison to that of intracellular liquid and do not take the orientation of real fibers into account. Single cables, in their interpretation, consist of a system of the cells' current generators (taken from the Beeler Reuter modified AP model [17]) connected through intracellular resistance R_x proportional to the cell's length. The distance between the neighboring cables is equal to the diameter of a cell, *d*. The interconnection between these cables is realized with space interval Δ by intracellular resistance R_y in both directions. Thus, interconnections between cables are much sparser then between cells in a cable. That and use of $R_x<R_y$ allows the reproduction of a given uniform anisotropy in simulated tissue. The mathematical description of this representation of 2D tissue is given in the appendix.

The authors of this approach considered that there are two major computational advantages of this method:

1. it significantly reduced matrix sizes, which must be inverted using conventional numerical approaches;
2. it makes parallelization of the computational algorithm possible.

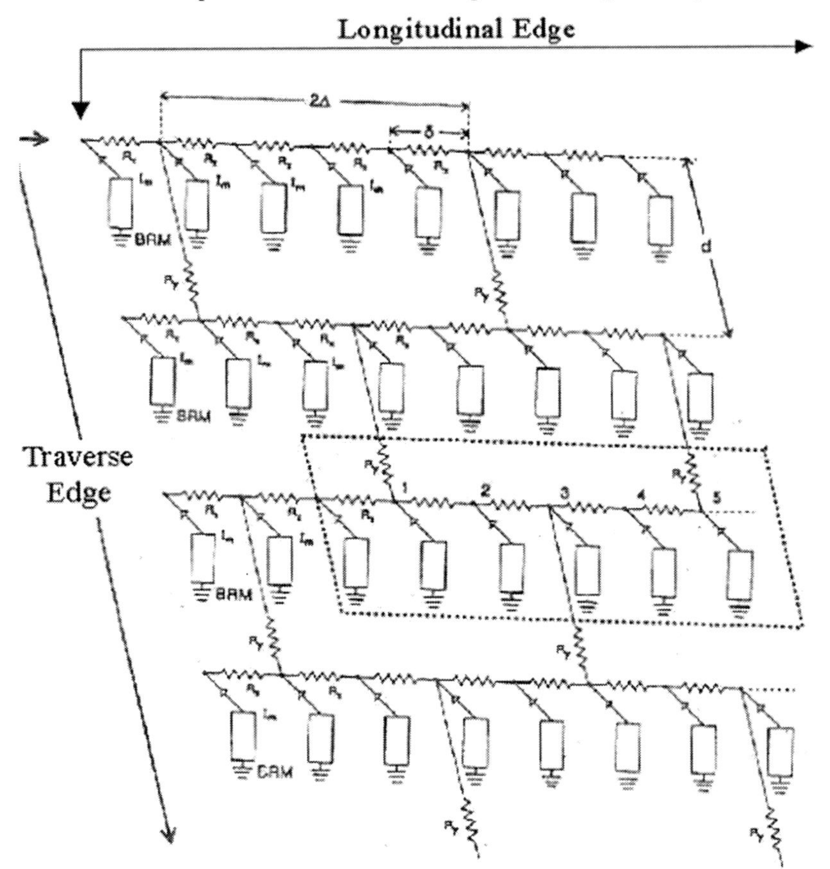

Fig. 3. Representation of heart muscle model as a system of parallel and interconnected cables. BRM is the abbreviation for the Beeler Reuter AP model modification [17].

The geometry of the 2D tissue model in which excitation-wave propagation was studied is shown in fig. 4. In particular, fig. 4 illustrates the case of excitation-wave propagation from a narrow path to an open space. It is easy to show that there exists a critical value of the pass width below which a conduction block appears.

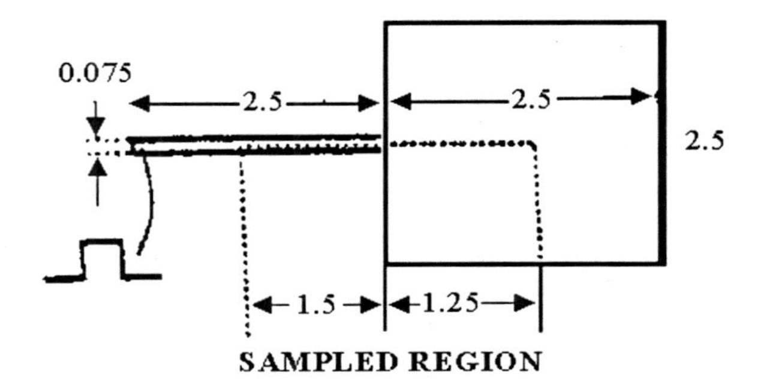

Fig. 4. Wave propagation from a narrow pass into an expanse of tissue.

Unfortunately, this method was not widely applied and deserves to be reconsidered for cardiac tissue with developed cellular Ca dynamics and a bidomain representation. The method is practical given the increasing power of modern computers and new numerical algorithms used for parallel computation of such problems. Some difficulties may arise in simulation of these problems for 3D tissue with transmural heterogeneity and variable fiber orientation.

8.4. Propagation of rectilinear front

Propagation of a solitary wave with a rectilinear front is shown in fig. 5. In normal cardiac tissue, the conduction velocity of a rectilinear front achieves the value $\theta \cong 50\ [\frac{\text{cm}}{s}]$. The wavelength is defined as:

$$\lambda = \theta\, APD \tag{12}$$

For APD=250 ms, and $\theta = 50\text{cm/s}$, the wavelength will be $\lambda = 12.5\text{cm}$.

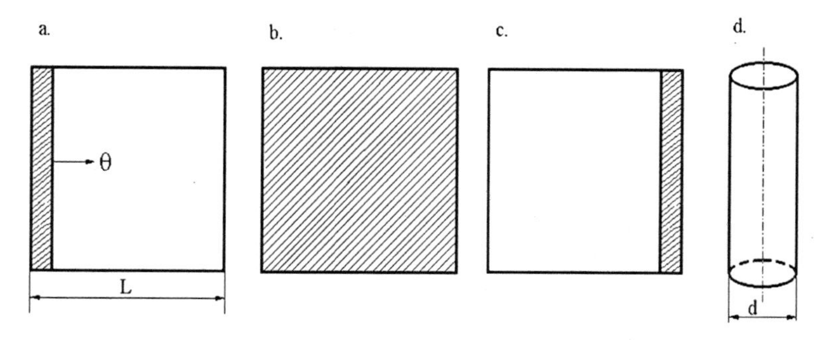

Fig. 5. Rectilinear wave propagation in homogeneous 2D tissue: a. Initiation of a propagating wave. b. wavelength is longer than tissue length. c. Wave exits tissue through a border. d. Cylindrical tissue shape with diameter $d = L$.

Propagation of a sequence of waves with rectilinear fronts in 2D tissue is the same as that in 1D fiber with restricted length (cases a, b, c in Fig. 5). The propagation of an excitation wave with a rectilinear front along a closed cylindrical 2D surface (case (d) in fig. 5) may be considered a set of propagating waves in a ring-shaped 1D tissue.

8.5. Propagation of wave with curvilinear front

Propagation of an initially rectilinear front in inhomogeneous tissue leads to its curvature, shown in fig. 6. The wave with a curvilinear front, after some transient period, propagates in this case without changing its shape.

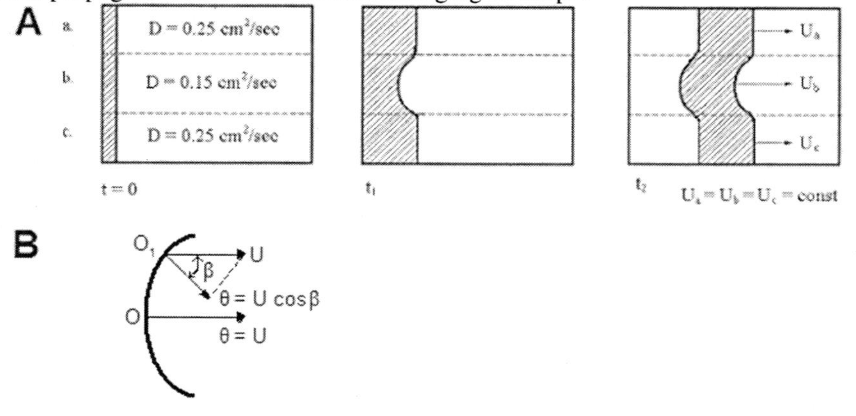

Fig. 6. A. Propagation in heterogeneous 2D tissue leads to the appearance of a curvilinear front. Heterogeneity is introduced using different diffusion values for tissue strips a, b, c ($D_1 = 0.25$ for strips a, c and $D_2 = 0.15$ for strip b). B. Relationship between the translational and normal conduction velocity components of a point on the curvilinear wavefront.

8.5.1. Circular wave as an example of wavefront with equal curvature, K, in each point on the wavefront.

The propagation of circular waves illustrated in Fig. 7 gives an example when $K = 1/Z$, where Z is the radius of circular wavefront. For a given time t, Z and K are constant at all points of the circular front.

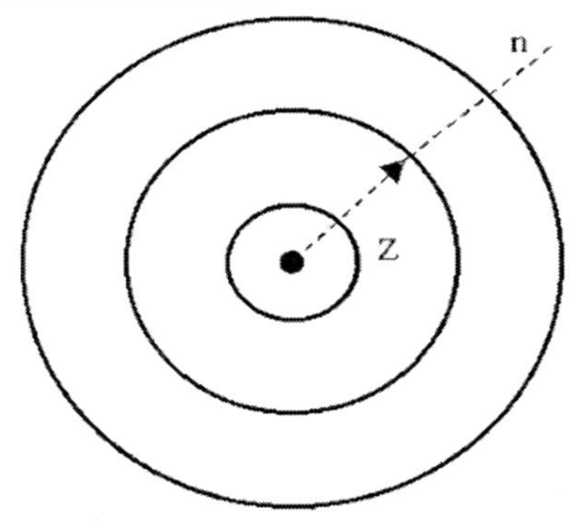

Fig. 7. Propagation of circular wavefronts with equal normal conduction velocity at each point of the front at fixed instants of time.

Normal propagation (see Fig. 8) of a circular wave is initiated by external stimulation applied to an $n \times n$ subgrid in the upper right corner of the whole tissue model. The number n must satisfy the source-sink conditions in order to obtain a propagating wave.

The states of tissue excitation are shown in fig. 8 at fixed propagation times. Due to diffusion properties, the original square-shaped excited tissue area is transformed into a propagating one-quarter wave portion with a circular front (Fig. 8B, left map).

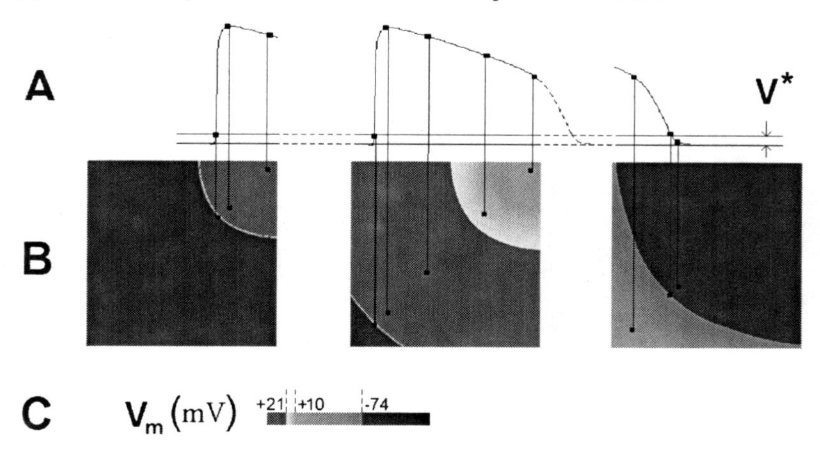

Fig. 8. Circular wave propagation initiated from the right upper corner of 2D tissue. A. Traces of the spatial distribution of membrane potential V_m along a diagonal emanating from the right upper corner of the tissue at different times after the initiation of wave propagation. The term $\mathbf{V}^*$ defines the front and tail of the propagating wave. B. Spatial maps of V_m in 2D tissue (6.4 cm x 6.4 cm) from which the spatial distributions of A are obtained. These results were obtain with the Chudin AP model [11] using parallel supercomputer NERSC IBM p575 POWER 5. C. The color code used in Fig. 8B.

The excited areas are indicated by multicolor regions that are *not* blue in Fig. 8B, center map and are defined as a cluster of nodes where voltage exceeds a certain threshold V^*. When the excitation wave propagates, this voltage level is reached twice: when the nodes enter (when $\frac{\partial V_m}{\partial t} > 0$) and exit (when $\frac{\partial V_m}{\partial t} < 0$) excitation. The narrow area of these nodes entering excitation and moving towards the unexcited region of the tissue grid is called the wavefront. The nodes going out of excitation and moving toward the recovering nodes form the wavetail.

8.6. Approaches for Spiral Wave Initiation in Computer Simulations

All approaches are based on the creation of a prematurely stimulated zone before the front or behind the tail of a propagating wave. Generally, the initiation of spiral waves in originally homogeneous excitable media requires the introduction of some temporal heterogeneity in the tissue. Gulko and Petrov [3] were the first who showed, in computer simulations, the initiation of spiral waves when some part of the tissue becomes temporarily unexcitable (Fig. 9), and when a premature stimulus S_2 is applied before the front (Fig. 10) and behind the tail (Fig. 11) of a basic propagated wave S_1.

The cross-hatched area in Fig. 9 is made temporarily unexcitable. In the left upper corner of the simulated tissue, a basic propagating wave S_1 is initiated with a circular front in time $t<t_1$. This wave reaches the border of the unexcitable segment, stops, and the cells in a portion of its front begin to repolarize and turn into the tail of the wave. A point q subsequently emerges, where the front and tail of the wave are

joined (by the time t_2). The S_1 wave then morphs into a single spiral wave with *bq* as its tail and *aq* as its front (t_2 and t_3 and further). Excitability in the S_2 area is then restored (after time t_3).and the spiral wave then continue to circulates in the fully excitable tissue model.

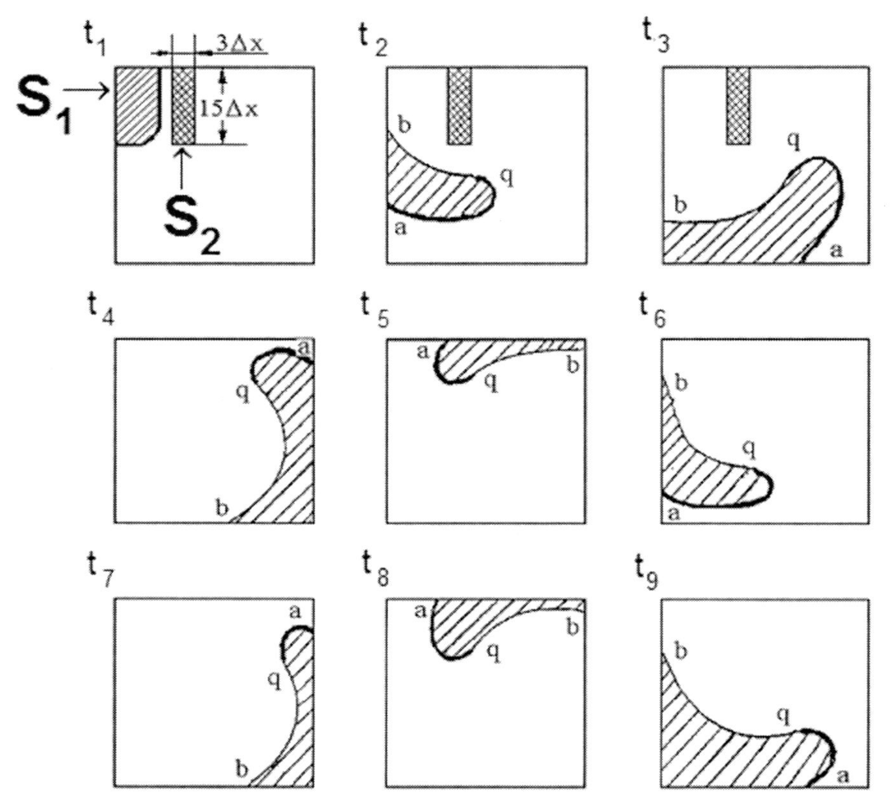

Fig. 9. The cells of a certain segment of cardiac excitable tissue (double shading) becomes temporarily unexcitable, remaining at the cell membrane rest potential. The size of the tissue model was 35x35 nodes with equal space steps $\Delta x = \Delta y = 0.05mm$ and the time step $\Delta t = 2.5ms$. The simplified model of cardiac cell proposed in [18] served in these computer simulations (performed using Hybrid Computer System HCS-100 [19]) as a point model.

The temporarily unexcitable region of tissue may be created by a number of other ways. For example, Fig. 10 illustrates how it is possible to obtain in computer experiments a spiral wave by applying an additional stimulus to some segment S_2 of tissue before the front of a propagating wave S_1. In this case, the size of a stimulated area and stimulus magnitudes must be small enough to not allow the origination of a propagating wave. Therefore, at the instant when primary wave S_1 approaches the segment S_2, all of its nodes transfer to the absolute refractory period of the repolarization phase and hence serve as temporarily unexcitable tissue for the arrived S_1 wave.

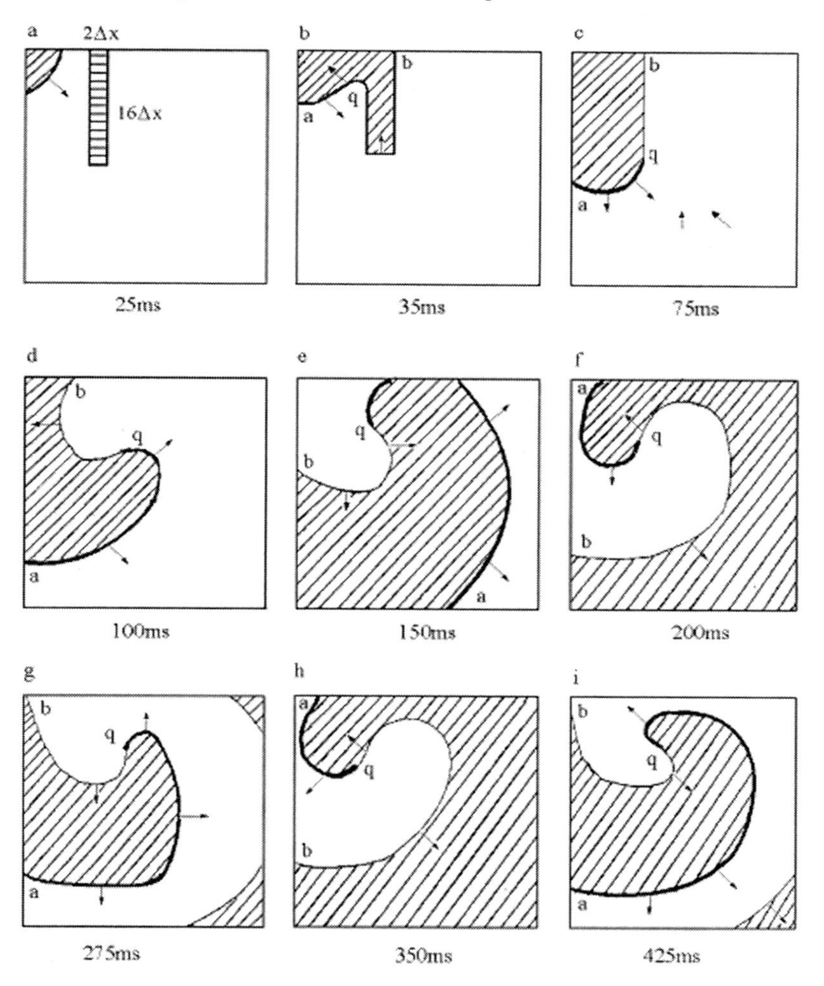

Fig.10. Premature excitation applied before the front of the wave. These results were obtained under the same conditions used for Fig. 9.

Another method to initiate reentry wave processes (see Fig. 11) is to create a zone of premature stimulation near the tail of the basic excitation wave S_1 caused by pacemaker activity. This prematurely stimulated area can generate an additional propagating wave only in the direction opposite to that of the wave S_1. Otherwise, the refractory properties of the S_1 area will present an insurmountable obstacle. The front of a premature wave of excitation directed toward a tail of a basic wave will stop and turn to repolarize, creating the conditions for forming the point *q*. Depending on the location of the premature stimulus area in relation to the border of a tissue, it is possible to generate single or double spiral waves. This method resembles, to some extent, the Wiener and Rosenblueth approach [20] to create solitary excitable wave propagation in 1D ring-shaped tissue by applying two consecutive stimuli to different points on the ring (also see chapter 7).

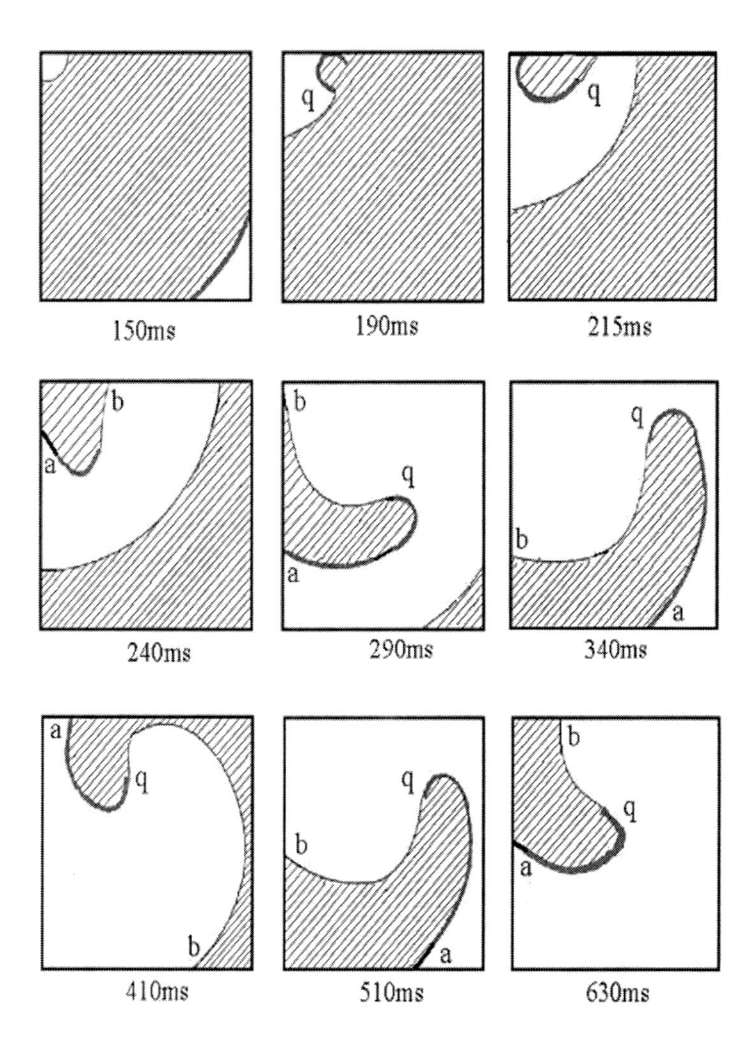

Fig. 11. Premature excitation at the tail of the propagating wave. The parameters and size of the tissue as well the type of used hybrid computer are the same as mentioned for the previous two figures.

It is necessary to note that these figures show the initiation of spiral waves when a previous propagating wave had a circular morphology. The location of a premature stimulus close to the tissue border is the reason why only single spiral waves were obtained. Double spiral waves can be initiated if the premature stimulus is applied in the middle of the tissue. The different approaches to initiating double and single spiral waves are illustrated in Fig. 12 and 13, respectively.

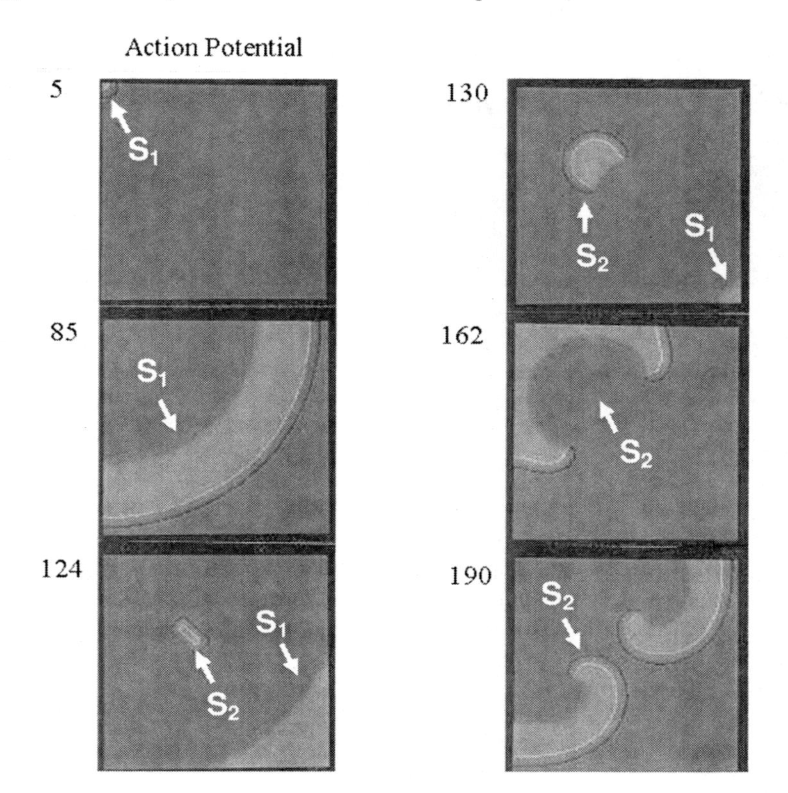

Fig. 12. Initiation of double spiral waves using a FitzHugh-Nagumo AP model with modified restitution properties [21] in a tissue model of 35 x 35 nodes. The simulations were performed using the massively parallel computer CM-2 introduced by the Thinking Machines Corporation. The numbers shown close to the upper left corners of the tissue indicate the time instances in ms related to different S_1 and S_2 waves locations.

The initiation of a spiral wave in initially uniform tissue by the method of cross field stimulation when S_1 and premature S_2 waves have rectilinear fronts is illustrated in Fig. 13.

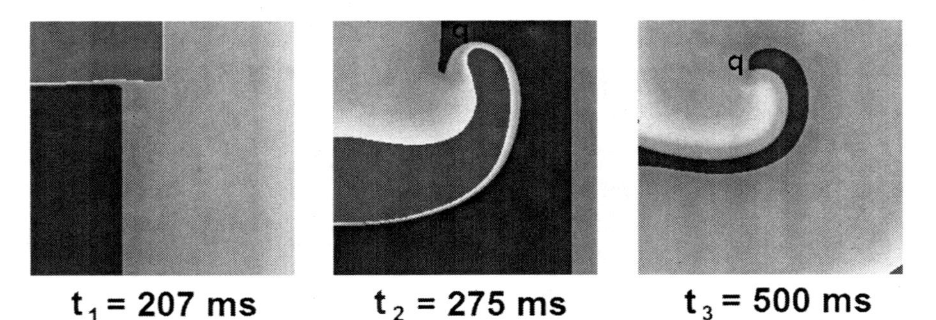

Fig. 13. Initiation of a single spiral wave by cross field stimulation in a tissue model of 256x256 nodes (measuring 6.4 cm x 6.4 cm) using the Chudin AP model [12]. Computer simulations were performed using the NERSC IBM p575 POWER 5 parallel supercomputer. The color code is the same used in Fig. 8B.

Comparison of different methods of initiating the spiral waves shows that characteristics of these waves do not depend on the specifics of the method. The method mostly affects the duration and behavior of the transient period during formation of a spiral wave. The location of the premature stimulation area in relation to the tissue border defines the type (single or double) of the obtained spiral wave.

8.7. Stationary and non-stationary spiral waves

The shapes of the cores (trajectory of the point q) for stationary and nonstationary propagation are shown in Fig. 14a and 14b, respectively. Fig. 15 illustrates the separation of the parameter space into regions of stationary and nonstationary propagation for 2D tissue composed of simplified FitzHugh-Nagumo type of AP models.

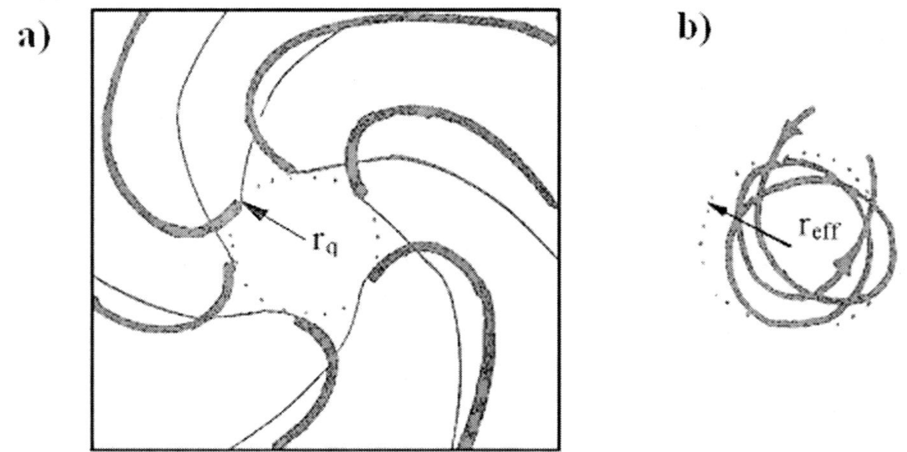

Fig. 14. a) Stationary spiral wave propagation with circular trajectory of the point q. Here it is shown that a comparatively small part of the tissue in the vicinity of the point q trajectory b) One of the possible point q trajectories during nonstationary propagation [5].

Stationary propagating spiral waves, as it follows from Fig. 14a, occur when the point q moves around a regular circle with constant angular velocity. All points must move with the same angular velocity on the front and tail of a spiral wave. By changing the parameters of the cell and tissue, it is possible to obtain nonstationary propagation. If nonstationarity is not strong, the trajectories of the point q do not extend beyond a circle of radius r_{eff} (Fig.14b).

The stationarity of wave circulation and values of r_q and $r_{q,eqv}$ depend on the parameters of tissue cells [5]. Corresponding results for a tissue model made up of simplified AP models are presented in Fig. 15.

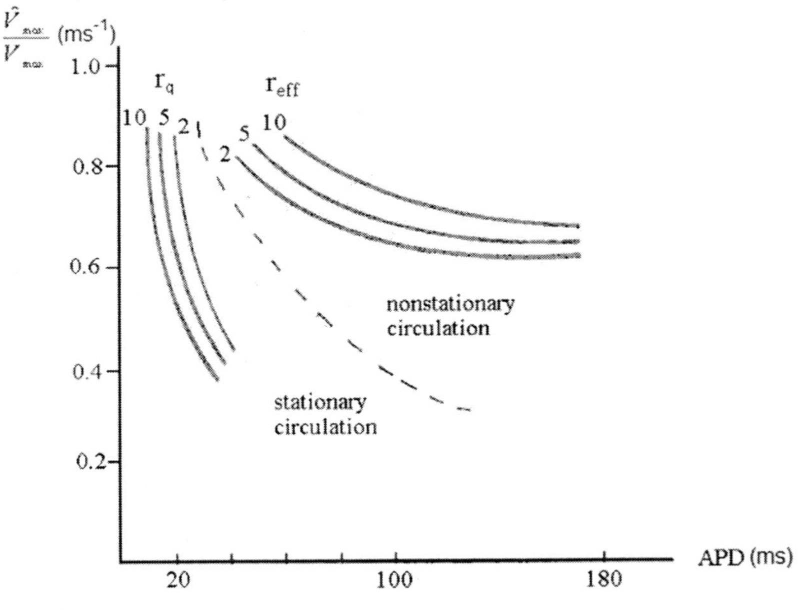

Fig. 15. Curves of different values of r_q and $r_{q.eff}$ in the parameter space (APD, $\dot{V}_{\max}/V_{\max}$) obtained for FitzHugh–Nagumo type AP model [5].

From this figure, it follows that increase of a cell's APD in tissue with a propagating spiral wave facilitates the transfer to nonstationary propagation. This explains why nonstationary spiral propagation is observed in tissues with Ca dynamics.

8.8. Curvature and Dispersion Relations

Experimentally (in course of physiological and computer simulation experiments) it was found that normal velocity of stationary spiral wave propagation at a given point on the front depends on curvature, K, at this point and the period, T, of spiral wave rotation. This means that there must exists a relationship:

$$\theta(s,t) = f\left[K(s,t), T(s,t)\right] \tag{13}$$

Where:

- θ – is the normal component of conduction velocity at the point *s* of the wavefront.
- K – is the curvature of the wavefront at the point *s*, which reflects the effect of diffusion currents in this point.
- T – is the elapsed time from the last excitation of the point *s*, which characterizes the distribution of cell recovery processes between successive waves.

The major task of the theory of stationary spiral wave propagation in excitable media and, in cardiac tissue in particular, is to find an explicit form of (13) and the major parameters of the cardiac cell and tissue, which determine *K* and *T*.

Unfortunately, these goals are not yet fully achieved. Significant contributions to the theory of the creation of stationary spiral wave propagation were made independently by Fife [22], Zykov [5], and Keener [23].

Here and in the following chapters, we present the most important elements of this theory, which, together with computer simulation experiments, help to explain, at least qualitatively, the properties of stationary spiral wave propagation in cardiac tissue. For our purposes, it is convenient to consider the relationship (13) for two extreme cases:

a) when the effect of cardiac cell gate variable recovery processes is finished before the next wavefront reaches the considered point, *S*, of a previous wavefront.
b) when the wavefront is close to rectilinear and the conduction velocity is dependent predominantly on gate variable recovery processes.

This approximate approach allows us to replace (13) by two relationships: the curvature relation, $\theta = f_1(K, T_\infty)$; and the dispersion relation, $\theta = f_2(0, T)$. Let us begin with the curvature relation, $\theta = f_1(K, T_\infty)$, which determines the component of conduction velocity when the period of spiral wave rotation is significantly higher than the time it takes cell processes to recover.

8.8.1. Curvature equation

Let us consider propagation in 2D isotropic tissue using (14):

$$\alpha\left(\frac{\partial^2 V_m}{\partial x^2} + \frac{\partial^2 V_m}{\partial y^2}\right) = C_m \frac{\partial V_m}{\partial t} + I_{stim} + I_{ion}(V_m, m) \tag{14}$$

$$\tau_m(V_m)\frac{\partial m}{\partial t} = m_\infty(V_m) - m$$

$$m = \{m_1, m_2, \ldots, m_k\}$$

$$\alpha = \frac{r}{2R_i}\left[\frac{1}{k\Omega}\right]; \alpha = \frac{1}{300} \div \frac{1}{800}\left[\frac{1}{k\Omega}\right]; \quad C_m = 1\frac{\mu F}{cm^2}$$

The Laplacian reflects the effect of local currents acting at the given point on the wavefront. They can be expressed through the curvature of the wavefront in this point. Indeed:

$$\frac{\partial^2 V_m}{\partial x^2}+\frac{\partial^2 V_m}{\partial y^2}=div(grad\ V_m)$$

From vector analysis follows:

$$divgradV_m=(\overline{n}, grad(gradV_m,\overline{n})+(gradV_m,\overline{n})div\overline{n} \tag{15}$$

Introducing curvilinear coordinate z, a read off along the line of current flow allows us to obtain:

$$(gradV_m,\overline{n})=-\frac{dV_m}{dz} \tag{16}$$

and

$$(\overline{n}, grad\,(grad\,V_m,\overline{n}))=\frac{\partial^2 V_m}{\partial z^2}$$

Taking into consideration that

$div\overline{n}=\frac{\partial n_x}{\partial x}+\frac{\partial n_y}{\partial y}$. Here, n_x and n_y are the projections of the unit normal vector to the point S on the x and y axes. Thus, $n_x = n\cos(\beta)$ and $n_y = -n\sin(\beta)$. So,

$$\frac{\partial n_x}{\partial x}=-\sin(\beta)\frac{\partial\beta}{\partial S}\frac{\partial S}{\partial x}$$

$$\frac{\partial n_x}{\partial x}=-\cos(\beta)\frac{\partial\beta}{\partial S}\frac{\partial S}{\partial y}$$

By definition, $K = \partial\beta/\partial S$. Here, β is the angle between the direction of vector n and the x-axis at the point S. Therefore, $\partial S/\partial x = \sin(\beta)$ and $\partial S/\partial y = \cos(\beta)$. Finally, we obtain

$$div(\overline{n})=-K\left(\sin^2(\beta)+\cos^2(\beta)\right)=-K$$

Substituting this result and (16) into the second part of (15), we get

$$\frac{\partial^2 V_m}{\partial x^2}+\frac{\partial^2 V_m}{\partial y^2}=\frac{\partial^2 V_m}{\partial z^2}+K\frac{\partial V_m}{\partial z} \tag{17}$$

The propagation equations now are:

$$\alpha\left(\frac{\partial^2 V_m}{\partial z^2}+K\frac{\partial V_m}{\partial z}\right)=c\frac{\partial V_m}{\partial t}+I(V_m,m)+I_{st}$$

$$\tau_m(V_m)\frac{\partial m}{\partial t}=m_\infty(V_m)-m \tag{18}$$

Let as introduce the following assumptions:

- K does not depend on z
- The wavefront propagates stationarily (θ = constant)

Changing the variable z to $\xi = z + \theta t$, we obtain:

$$\alpha \frac{\partial^2 V_m}{\partial \xi^2} + \alpha K \frac{\partial V_m}{\partial \xi} - c\theta \frac{\partial V_m}{\partial \xi} = I(V_m, m) + I_{st}\,;\quad \theta \tau_m(V_m) \frac{\partial m}{\partial \xi} = m_\infty(V_m) - m$$

or

$$\alpha \frac{\partial^2 V_m}{\partial \xi^2} + (\alpha K - c\theta) \frac{\partial V_m}{\partial \xi} = I(V_m, m) + I_{st} \tag{19}$$

$$\theta \tau_m(V_m) \frac{\partial m}{\partial \xi} = m_\infty - m$$

Let us consider the coefficient before $\frac{\partial V_m}{\partial \xi}$ in (17). Designating it by $\gamma = K\alpha - c\theta$

For $K = 0$, $\theta = \theta_{RL}$. So, $\gamma = -c\theta_{RL}$

$$\theta = \theta_{RL} + \frac{\alpha}{c} K \;;\; \frac{\alpha}{c} = D. \tag{20}$$

Here $K < 0$ for convex front and $K > 0$ for concave front.

Equation (20) represents the linear dependence between the normal velocity at a given point of the propagating wavefront and its curvature at that point. That is certainly an estimate. In reality, as shown in Fig. 16, only a part of this relation is linear and critical curvature is reached at some small, but non-zero, level of wavefront velocity.

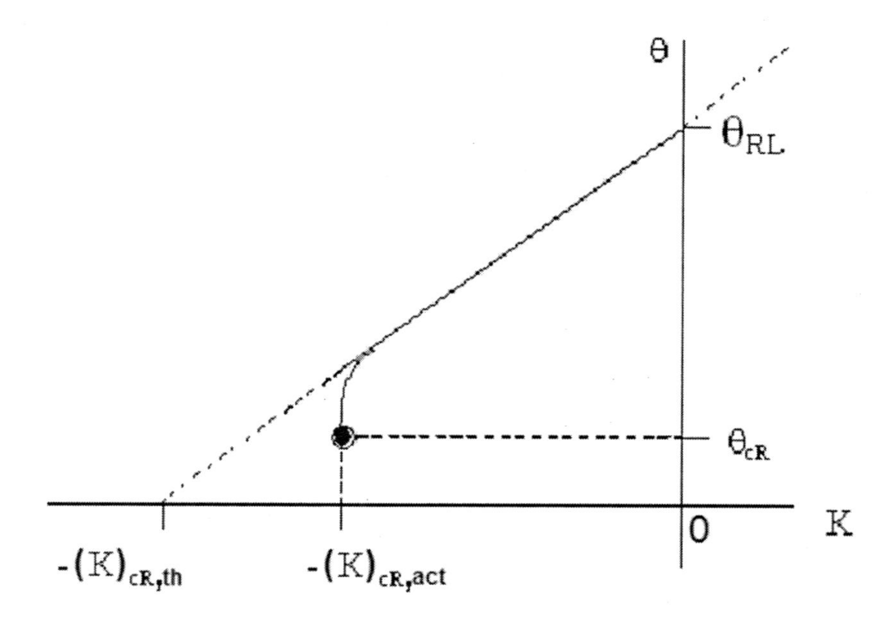

Fig. 16. The theoretical (th) and actual (act) curvature relations.

The curvature equation is an estimate. For a different AP model, it is necessary to introduce a correction coefficient, *v*. The curvature equation now has the form:

$\theta = \theta_{RL} + \nu \frac{\alpha}{c} K$, where $\theta_{RL} = f_2(T)$.

8.8.2. Dispersion Relation

The dispersion relation reflects the dependence of conduction velocity on the dispersion of cell recovery processes in time *T* elapsed from the previous excitation $\theta = \theta_{RL} = f_2(0,T)$ when curvature is equal to zero. Recovery processes distribution in time and space may be measured during computer simulations.

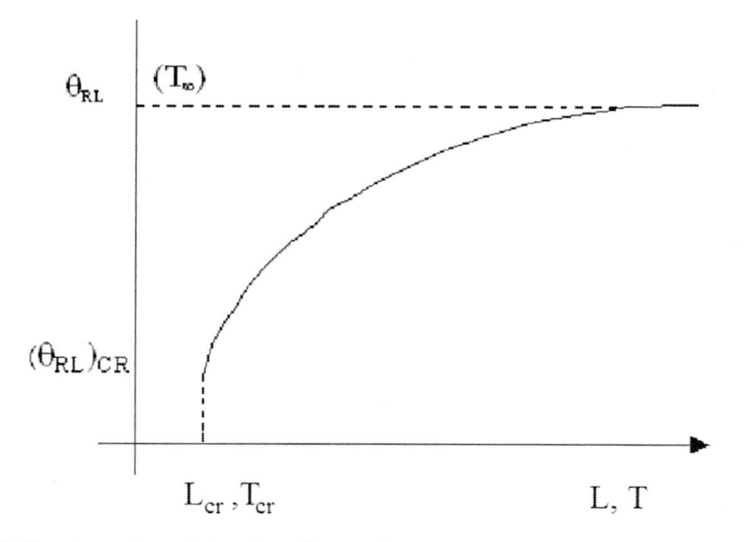

Fig. 17. The dispersion relation. L – distance between two stationary spiral wavefronts. T- time elapsed from the previous front of a stationary wave.

The character of the dispersion relation shown in Fig. 17 is obtained from general considerations. The details depend on the cellular characteristics of the tissue, and there is not yet enough data available to present specifics.

8.8.3. The effect of diffusion coefficient

As it is known, the diffusion coefficient is defined as $D = \frac{\alpha}{C}$. In order to introduce the diffusion coefficient in the propagation equation explicitly, let us divide both sides of that equation by *C*. We obtain:

$$D\left(\frac{\partial^2 V_m}{\partial x^2} + \frac{\partial^2 V_m}{\partial y^2}\right) = \frac{\partial V_m}{\partial t} + \frac{I_{st} + \sum I_{ion}(V_m, m)}{C} \qquad (21)$$

Introducing the substitution:

$$\bar{x} = \frac{x}{\sqrt{D}} \text{ and } \bar{y} = \frac{y}{\sqrt{D}} \tag{22}$$

We finally get (21) in the form:

$$\frac{\partial^2 V_m}{\partial \bar{x}^2} + \frac{\partial^2 V_m}{\partial \bar{y}^2} = \frac{\partial V_m}{\partial t} + \frac{I_{st} + \sum I_{ion}(V_m, m)}{C} \tag{23}$$

We can consider (23) as a special case of (21) when $D = 1$. This allows us to conclude that the diffusion coefficient affects only the space characteristics of the solution. Each parameter, with dimensions including space, can be represented as:

$$P[x] = \sqrt{D}\ (P)_{D=1}$$

For the example for conduction velocity in 1D tissue:

$$\theta_x = \frac{dx}{dt} = \sqrt{D}\frac{d\bar{x}}{dt} = \sqrt{D}\,(\theta)_{D=1}$$

8.8.4. Direct computer simulation approach to find the curvature relation

Here we present the results from determining the curvature relation for isotropic and anisotropic cardiac tissue using a computer simulation method proposed in [24]. The results presented in Fig. 18 are in good agreement with those obtained in [5] for isotropic tissue using the FitzHugh-Nagumo AP model with standard parameters.

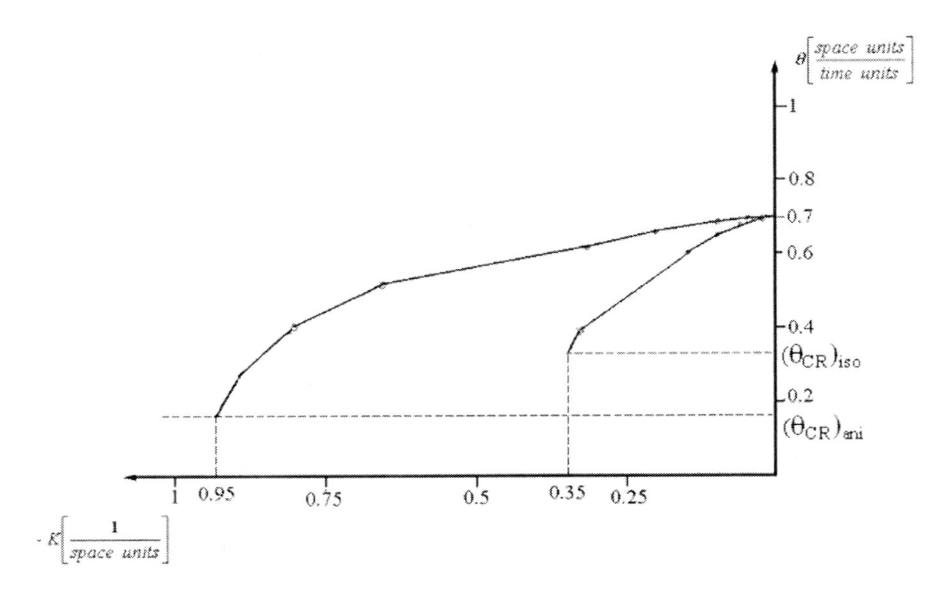

Fig. 18. Curvature relationships for isotropic and anisotropic tissue with the FitzHugh-Nagumo AP model.

8.9. Appendix: 2D Myocardium Modeled as Interconnected Cables

For an isolated j fiber:

$$\frac{a}{2r_i}\frac{\partial^2 V^j(x,t)}{\partial x^2} = C_m \frac{\partial V^j(x,t)}{\partial t} + \sum I_{ion}(x,t) + I_{St} \tag{24}$$

Discretization of the time t gives:

$$\frac{\partial V^j}{\partial t} = \frac{V^j(t_{i+1},x) - V^j(t_i,x)}{t_{i+1} - t_i}$$

Let us introduce the following definitions:

$$\Delta t_i = t_{i+1} - t_i$$

$$\frac{a\Delta t_i}{2r_i} = D_i = D$$

$$-V^j(x,t_i) + \left[\sum I_{ion}(x,t_i) + I_{St}\right]\Delta t_i = F(x,t_i)$$

$$D\frac{\partial^2 V^j(x,t_{i+1})}{\partial x^2} - V^j(x,t_{i+1}) = F^j(x,t_i) \tag{25}$$

The solution

$$V^j(x,t_{i+1}) = V_h^j + V_p$$

$$D\frac{\partial^2 V^j(x,t_{i+1})}{\partial x^2} - V^j(x,t_{i+1}) = 0$$

$$V_h^j = C_1 e^{\frac{x}{\sqrt{D}}} + C_2 e^{-\frac{x}{\sqrt{D}}}; \qquad V_p^j(x,t_i)$$

For cable with junctions

$$V_k = V_{kh} + V_{kp} = V_{kh} + V_p \quad \text{for } \hat{x}_K < x < \hat{x}_{K+1}$$

$$V_{kh} = b_{k1} e^{\frac{x}{\sqrt{D}}} + b_{k2} e^{-\frac{x}{\sqrt{D}}} + V_p$$

b_{k1} and b_{k2} can be obtained from the junction conditions:

$$I_{k-0} + I_{k+0} + I_{R_n} = 0$$

$$I_{R_n} = \frac{V^{j+1}(\hat{x}_k) - V^j(\hat{x}_k)}{R_n}$$

$V(x)$ - continuous

$$V_p^j(x) = \frac{1}{2\sqrt{D}}\left[e^{\frac{x}{\sqrt{D}}} \int_0^x e^{-\frac{x}{\sqrt{D}}} F(x)dx - e^{-\frac{x}{\sqrt{D}}} \int_0^x e^{\frac{x}{\sqrt{D}}} F(x)dx \right]$$

Results obtained with a two-dimensional spiral wave [16] are presented in fig. 19.

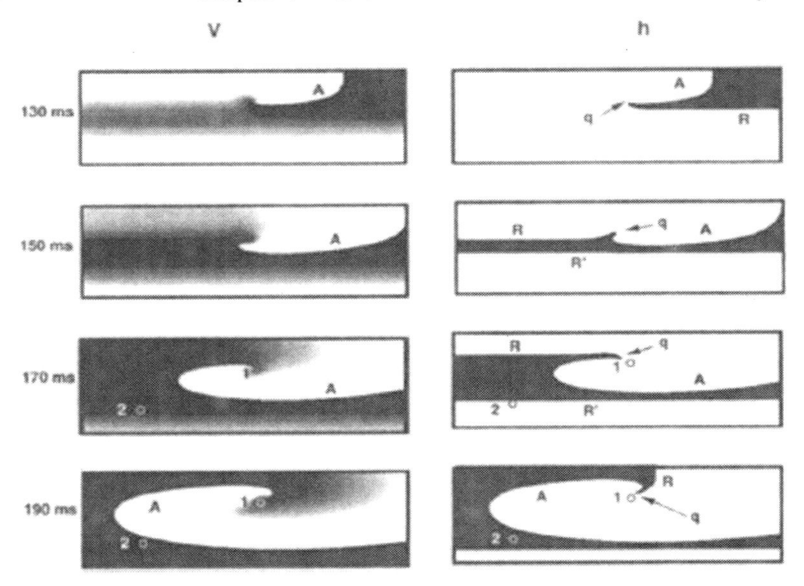

Fig. 19. Initiation of two-dimensional reentry. AP distribution (V-map, left column) and distributions of recovery from inactivation (h-map, right column) at 130-190 ms after application of stimulus S_1. Stimulus S_2 occurring 110 ms after S_1. A = activation front due to S_2, R = recovery front due to S_2; R′ = recovery front due to S_1; q = point of junction between the A and R fronts. Figure is an excerpt from [16].

Results in fig. 19 were obtained under the following conditions to achieve reentry:

- AR (anisotropy ratio) = 4:1 and
- time constants, τ_d and τ_f, in expression for I_{si} decreased by a factor of 8 to obtain the shortening of APD and ARP (absolute refractory period).

8.10. References

1. Winfree, A.T., *Mechanisms of cardiac fibrillation - Reply.* Science, 1995. **270**: 1224-1225.
2. Karma, A., *Electrical alternans and spiral wave breakup in cardiac tissue.* Chaos, 1994. **4**: 461-472.
3. Gulko, F.B. and A.A. Petrov, *Mechanisms of the formation of closed pathways of conduction in excitable media.* Biofizika (USSR), 1972. **17**: 261-270.
4. Allessie, M.A., F.I.M. Bonke, and F.J.C. Schopman, *Circus movement in rabbit atrial muscle as a mechanism of tachycardia.* Circ.Res., 1973. **33**: 54-77.
5. Zykov, V.S., *Simulation of Wave Process in Excitable Media.* Nonlinear science: theory and applications, ed. A.V. Holden. 1987, Manchester and New York: Manchester University Press.
6. Tyson, J.J. and J.P. Keener, *Singular perturbation theory of traveling waves in excitable media.* Physica D, 1988. **32**: 327-361.
7. Keener, J.P. and A.V. Panfilov, *The effects of geometry and fibre orientation on propagation and extracellular potentials in myocardium*, in *Computational Biology of the Heart*, A.V. Panfilov and J.P. Keener, Editors. 1997, John Wiley & Sons: New York, NY. p. 235-258.
8. Henriquez, C.S., *Simulating the electrical behavior of cardiac tissue using the bidomain model.* Crit Rev Biomed Eng, 1993. **21**: 1-77.

9. Neu, J.C. and W. Krassowska, *Homogenization of syncytial tissues.* Crit Rev Biomed Eng, 1993. **21**: 137-199.
10. Roth, B.J., *How the anisotropy of the intracellular and extracellular conductivities influences stimulation of cardiac muscle.* J Math Biol, 1992. **30**: 633-646.
11. Luo, C.H. and Y. Rudy, *A dynamic model of the cardiac ventricular action potential. I. Simulations of ionic currents and concentration changes.* Circ Res, 1994. **74**: 1071-1096.
12. Chudin, E., J. Goldhaber, A. Garfinkel, J. Weiss, and B. Kogan, *Intracellular Ca(2+) dynamics and the stability of ventricular tachycardia.* Biophys J, 1999. **77**: 2930-2941.
13. Huffaker, R.B., R. Samade, J.N. Weiss, and B. Kogan, *Tachycardia-induced early afterdepolarizations: Insights into potential ionic mechanisms from computer simulations.* Comput Biol Med, 2008. **38**: 1140-1151.
14. Rudy, Y. and W. Quan, *A model study of the effects of the discrete cellular structure on electrical propagation in cardiac tissue.* Circ.Res., 1987. **61**: 815-823.
15. Leon, L.J., F.A. Roberge, and A. Vinet, *Simulation of two-dimensional anisotropic cardiac reentry: effects of the wavelength on the reentry characteristics.* Ann Biomed Eng, 1994. **22**: 592-609.
16. Leon, L.J. and F.A. Roberge, *Structural complexity effects on transverse propagation in a two-dimensional model of myocardium.* IEEE Trans Biomed Eng, 1991. **38**: 997-1009.
17. Drouhard, J.P. and F.A. Roberge, *Revised formulation of the Hodgkin-Huxley representation of the sodium current in cardiac cells.* Comput Biomed Res, 1987. **20**: 333-350.
18. Gulko, F.B. and A.A. Petrov, *On a mathematical model of excitation processes in the Purkinje fiber.* Biofizika (USSR), 1970. **15**: 513-520.
19. Kogan, B. and P. Vrbavatz, *General structure of the hybrid computer system HCS-100.* 1974, Moscow Institute of Control Science: Moscow, Russia.
20. Wiener, N. and A. Rosenblueth, *The mathematical formulation of the problem of conduction of impulses in a network of connected excitable elements, specifically in cardiac muscle.* Arch. Inst. Cardiol. Mexico, 1946. **16**: 205-265.
21. Kogan, B.Y., W.J. Karplus, B.S. Billet, A.T. Pang, H.S. Karagueuzian, and S.S. Khan, *The simplified Fitzhugh-Nagumo model with action potential duration restitution: effects on 2D-wave propagation.* Physica D, 1991. **50**: 327-340.
22. Fife, P.C., *Mathematical Aspects of Reacting and Diffusing Systems.* 1979, Berlin, Germany: Springer-Verlag.
23. Keener, J.P., *An eikonal-curvature equation for action potential propagation in myocardium.* J Math Biol, 1991. **29**: 629-651.
24. Kogan, B.Y., W.J. Karplus, B.S. Billet, and W. Stevenson, *Excitation wave propagation within narrow pathways: geometric configurations facilitating unidirectional block and reentry.* Physica D, 1992. **59**: 275-296.

Chapter 9. Theory and Simulation of Stationary Wave Propagation

Stationary spiral wave phenomena, in simulated myocardium consisting of AP models, and their associated characteristics (such as rotational angular velocity, core radius, and wavefront morphology) are topics of major theoretical and practical interest. These topics will be discussed in this chapter within the framework of findings from Zykov [1], which are valid under the following assumptions:

a. an unrestricted domain of spiral wave propagation
b. only one spiral wave is initiated
c. spiral wave rotation is stationary and the resultant period of circulation, $T = 2\pi/\omega$, is constant.

Various examples of stationary and nonstationary spiral wave propagation that are obtained from computer simulations are presented here. These examples serve to illustrate several theoretical results, demonstrate the effect of different parameters in AP and myocardium models on spiral wave initiation and propagation, and highlight specific features of intracellular Ca dynamics incorporated in recent AP models. In order to facilitate an understanding of these findings, a description of the kinematics of stationary wave propagation and related relationships is presented in the following section.

9.1. Kinematics of Stationary Spiral Wave Propagation

The kinematic characteristics of a stationary spiral wave can be derived using Fig. 1, which summarizes a simulation result shown in the previous chapter.

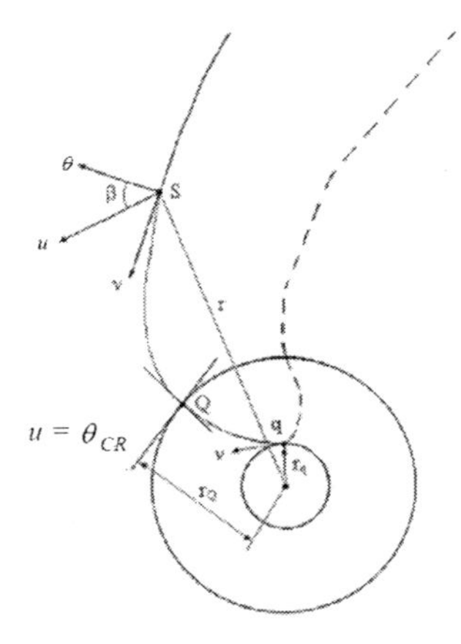

Fig. 1. Geometry of a stationary propagating wave. In this figure, r_q is the radius from point q to the edge of the rotating spiral wave, s is the location of a point on the spiral wave in reference to point q or Q. When measured in reference to q, $s = 0$ at point q, $s > 0$ for points on the wavefront, and $s < 0$ for points on the wavetail. If measured in reference to Q, then $s = 0$ at point Q and $s > 0$ for all points on the wavefront where $r > r_Q$.

B.Ja. Kogan, *Introduction to Computational Cardiology: Mathematical Modeling and Computer Simulation*, DOI 10.1007/978-0-387-76686-7_9,

Fig. 1 provides the means to derive several mathematical relationships. In the case of a stationary spiral wave, it will rotate around the center of the core as a solid body with a constant angular velocity, ω. The linear conduction velocity, u, for each point s on the wavefront will then be given by $u = \omega r$, where r is the radius of rotation from point q. The projections θ (normal to the point s) and v (tangent to the point s) of the velocity u are given by the expressions below

$$\theta = u\cos\beta \text{ and } v = u\sin\beta \tag{1}$$

where β is the angle between u and the unit normal. It is evident that u and its components vary with respect to r and wavefront curvature. Thus, in point q ($s = 0$):

$$u_q = \omega r_q = v_q \text{ and } \theta_q = 0. \tag{2}$$

Stationary spiral waves also have the property that their curvature in an unrestricted domain of propagation tends to zero and θ tends to θ_{RL} as the point s becomes infinitely distant from the point q. In this circumstance, a point Q on the wavefront always exists for which the following expression is true:

$$u_Q = \theta_Q = \theta_{CR} = \omega r_Q . \tag{3}$$

Combining (2) and (3) produces

$$r_q = r_Q \frac{v_q}{\theta_{CR}} . \tag{4}$$

Another characteristic of the spiral wave is the curvature of a curvilinear front, K, and this may be defined via several relations:

$$K(s) = \frac{\partial \beta}{\partial s},\ K(\rho) = \frac{1}{\rho},\text{ and } K(x, y) = \frac{\frac{d^2 y}{dx^2}}{\sqrt{\left[1 + (\frac{dy}{dx})^2\right]^3}} .$$

During stationary propagation, K is time independent.

9.1.1. Natural Equations and the Morphology of a Stationary Spiral Wavefront

Fig. 1 indicates that two characteristic points on the wavefront exist: point Q, where the velocities are described by the expressions $v = 0$ and $u = \theta = \theta_{CR}$; and point q, where the analogous expressions for velocity are $u = \overline{\tau} v$ ($\overline{\tau}$ is the unit tangent vector) and $\theta = 0$.

It is worthwhile to introduce a curvilinear system of coordinates in relation to the wavefront. The length of the arc s along the wavefront can be measured either from the point q or Q. In the former case, $s = 0$ at the tip of the spiral wave and s is considered positive along the wavefront and negative along the wavetail. For the second case, $s = 0$ at the point Q and s is positive when $r > r_Q$ and negative when $r < r_Q$. At each point s, the velocities θ and v can be expressed, as an alternative to (1), as scalar products: $\theta = (u, \overline{n})$ and $v = (u, \overline{\tau})$, where $\overline{n}$ is the unit normal vector.

By utilizing the Frenet formulas from differential geometry (see the Appendix) and some basic rules of vector analysis (see [1] for details), one can finally transform these expressions into the form

$$\frac{d\theta}{ds} + K(s)v(s) = \omega \tag{5}$$

$$\frac{dv}{ds} - K(s)\theta(s) = 0\,, \tag{6}$$

with the boundary conditions $\theta(s)\big|_{s=0} = \theta_{CR}$ and $v(s)\big|_{s=0} = 0$ and with $s = 0$ at the point Q. The equation set of (5) and (6) constitute the Natural Equations.

If the relation $u^2(s) = \theta^2(s) + v^2(s)$ is taken into consideration, then it is possible to obtain the following relation (see Appendix for details) from (5) and (6):

$$K^2(s) = \frac{(\omega - \frac{d\theta(s)}{ds})^2}{2\omega \int_0^s \theta(\xi)d\xi - \theta^2(s) + u^2(0)}\,. \tag{7}$$

In (7), the parameters ω and $u(0)$ are constant. This expression determines the curvature of the wavefront as a function of $\theta(s)$, namely $K(s) = f(\theta(s))$. This functional dependence of $K(s)$ may be determined either by computer simulation [2] or its first approximation as a linear curvature expression (see Chapter 8).

9.1.2. Estimates of the Stationary Spiral Wave Angular Velocity ω and the Radius r_q

The Natural Equations (5) and (6), along with the boundary conditions $K(0) = K_{CR}$ and $v(0) = 0$ (with $s = 0$ at the point Q) and the replacement of θ in (5) and (6) by the linear curvature equation $\theta(s) = \theta_{RL} + v_1 DK(s)$, give

$$\frac{dv}{ds} = K(s)(\theta_{RL} + D_1 K(s)) \quad (v(0) = 0) \tag{8}$$

$$D_1 \frac{dK}{ds} = \omega - K(s)v(s) \qquad (K(0) = K_{CR}), \tag{9}$$

where $D_1 = v_1 DK(s)$.

In the above set of equations, it is necessary to select a value of the parameter ω such that $K > 0$ when $s \geq 0$ and $K \to 0$ when $s \to \infty$. It is straightforward to demonstrate that ω does not depend on D_1. In order to show this, the following variables are rescaled: $\tilde{K} = K\sqrt{D_1}$; $\tilde{v} = v/\sqrt{D_1}$; $\tilde{\theta} = \theta/\sqrt{D_1}$; and $\tilde{s} = s/\sqrt{D_1}$. Substitution of these rescaled variables in (5) and (6) does not alter the form of the Natural Equations. It becomes evident that ω is dependent on θ_{RL} and K_{CR}, but not D_1.

Therefore, using the dimensional considerations proposed in [1], ω can be expressed in the form

$$\omega = -\theta_{RL} K_{CR} A\left(-\frac{K_{CR} D_1}{\theta_{RL}}\right). \tag{10}$$

In (10), the negative sign indicates that $K < 0$ and the function $A(-K_{CR}D_1/\theta_{RL})$ is obtained from numerical solutions of the system composed of (8) and (9) with different values of θ_{RL} and constant values of $K_{CR} = -1$ (cm^{-1}) and $D_1 = 1$ (cm^2/s).

Estimates for r_Q and r_q can be found from the elementary kinematic properties of a stationary wave:

$$r_Q = \frac{\theta_{CR}}{\omega} = \frac{\theta_{RL} + D_1 K_{CR}}{\omega} \text{ and } r_q = r_Q \frac{v_q}{\theta_{CR}}. \tag{11}$$

9.2. Propagation in Restricted Myocardium

Consider a tissue with a circular form, with finite radius r_B (see Fig. 2), and a boundary condition $(\partial V/\partial n)_B = 0$. In this case, the Natural Equations (5) and (6) are valid for this particular geometry. If we measure the arc length, s, from the point Q, then the boundary conditions for (5) and (6) also remain unchanged. Therefore, the Normal Equations are

$$\frac{d\theta(s)}{ds} + K(s)v(s) = \omega \text{ and } \frac{d\,v(s)}{ds} - K(s)\theta(s) = 0 \quad ,$$

with the initial conditions $v(0) = 0$ and $\theta(0) = \theta_{CR}$. In addition, the relation $\theta = \theta_{RL} + D_1K(s)$ holds when $K \geq K_{CR}$. Since $u_B = \omega r_B = \theta_B$ and $v_B = 0$ for all points on the myocardial border, the boundary conditions in this region will be $v(s_B) = 0$ and $\theta(s_B) = \omega r_B = \theta_{RL} + D_1K(s_B)$. However, $K(s) = 0$ when $s = s_B$; thus, the boundary condition for $\theta(s_B)$ is simply equal to θ_{RL} and $\omega = \theta_{RL}/r_b$.

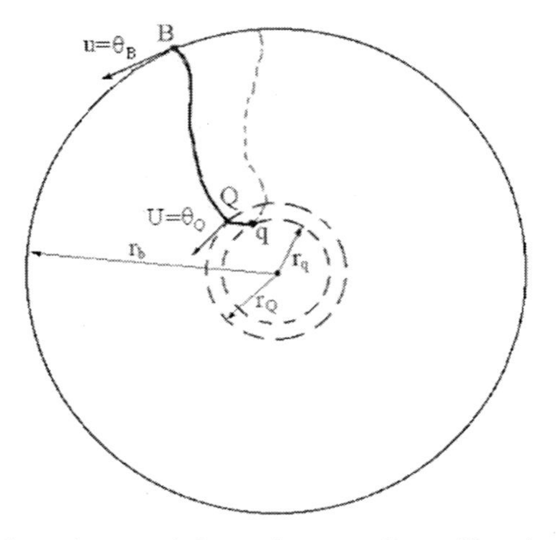

Fig. 2. Propagation of a stationary spiral wave in myocardium with a circular configuration and a radius r_b.

An inverse relationship exists between the radius r_b of a circular area of the myocardium and ω; however, it is true only for comparatively large r_b, for which the period of circulation is sufficiently long that the effects of recovery processes can be neglected. Computer simulation studies illustrate that decreasing the circular area of the myocardium leads to significant alteration of the spiral wave morphology near the core and smaller lengths for r_Q and r_q. Taken together, these findings show that myocardium configurations, as well as their sizes, play a significant role in stationary propagation. As an example, a square myocardium configuration is theoretically incapable of producing stationary spiral wave propagation, since the distances from the center of circulation to the borders are not equal in all directions.

9.3. Propagation in Unrestricted Myocardium with a Central Hole

Spiral wave propagation around a hole (with radius r_h) in the myocardium can occur as one of two cases (see Fig. 3): (1) when $r_h > r_Q$, (where r_Q corresponds to myocardium without a hole); and (2) when $r_h < r_Q$. It is possible to utilize equations (8) and (9), with the linear curvature expression, to analyze the first case. As a consequence, the following relations hold as the distance s becomes infinitely far from the hole: $K(s \to\infty) = 0$ and $\theta(s \to\infty) = \theta_{RL}$. The boundary conditions along the wave-hole interface, at the intersection point s_h, are obtained from the orthogonality condition: $v(s_h) = 0$ and $\theta(K(s_h)) = \omega r_h$.

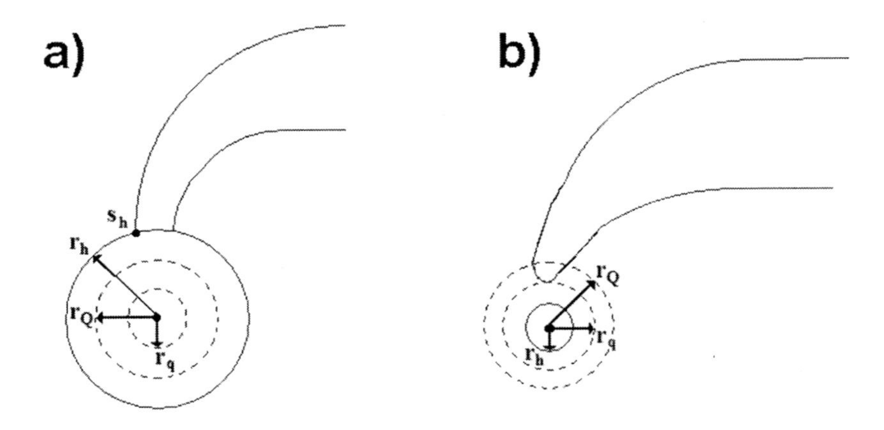

Fig. 3. Two cases of stationary propagation of a spiral wave in myocardium with an unrestricted size and a central hole: (*a*) when $r_h > r_Q$ and (*b*) $r_h < r_Q$.

In the second case (see Fig. 3*b*), stationary spiral wave propagation around the hole is possible if a new core of rotation arises during the formation of the spiral wave, with r_q and r_Q both exceeding r_h.

9.4. Propagation in 2D Myocardium with Simplified AP Models

9.4.1. General Considerations

The following examples presented here were selected in order to illustrate several propagation characteristics: (a) the effects of characteristics within simplified AP models on spiral wave initiation in 2D myocardium (after application of an appropriate premature stimulation); (b) the method of fitting APD restitution curves, obtained in real myocardium, into simplified AP models; and (c) distinctions between the morphologies of spiral waves in actual myocardium and in computer simulations with simplified AP models. Numerous publications regarding spiral wave initiation and propagation in simulated myocardium, which consist of first generation AP models, exist in the literature [3-5]. All these studies were primarily concerned with proving the so-called APD restitution hypothesis, which attempts to explain the drifting core of a meandering spiral wave and its subsequent breakup into multiple spiral waves. However, none of these investigations resulted in a formulation of the stability criteria for stationary spiral wave propagation in a 2D model of myocardium (in a similar manner to the Courtemanche *et al* study [6] in a 1D ring of Beeler-Reuter AP models). Nevertheless, simulation results obtained using first generation AP models demonstrate several important concepts and are included in this section. Later, findings from simulations of spiral wave propagation in 2D myocardium composed of second generation AP models are presented. These latter studies present qualitatively different spiral wave propagation and permitted the investigation of new phenomena, such as Ca and APD alternans during high pacing rates [7], spiral wave regeneration in myocardium with a restricted size [8], and the formation of EAD [9] and DAD [10] clusters during spiral wave rotation.

9.4.2. Simulations Using the FitzHugh-Nagumo Model

The results of computer simulations of stationary spiral waves in uniform myocardium that are presented here utilized the Pushchino modification of the FitzHugh-Nagumo (FHN) AP models (see details in Chapter 5), which were adopted in order to reproduce the cardiac AP. The mathematical model for isotropic myocardium ($\alpha_x = \alpha_y = \alpha$) can be represented in the following dimensionless form:

$$\frac{\partial \overline{V}_m}{\partial \bar{t}} = \left(\frac{\partial^2 \overline{V}_m}{\partial \bar{x}^2} + \frac{\partial^2 \overline{V}_m}{\partial \bar{y}^2}\right) + \overline{F}(\overline{V}_m) - \bar{I} + \bar{I}_{st} \tag{12}$$

$$\frac{\partial \bar{I}}{\partial \bar{t}} = \varepsilon(\bar{f}(\overline{V}_m) - \bar{I}) \tag{13}$$

Equations (12) and (13) consist of several dimensionless variables: $\overline{V}_m = V_m / V_{m,\max}$; $\bar{I} = I/I_{\max}$; $\bar{t} = t/\tau_m$; $\varepsilon(\overline{V}_m) = \tau_m/\tau_I$; and $\bar{y} = y/\lambda$. The functions $\overline{F} = F(V_m)/I_{\max}$ and $\bar{I} = I / I_{\max}$ represent the total inward and outward membrane currents (in dimensionless form), respectively, and they are represented using a piece-wise linear approximation. The function $\bar{f}(\overline{V}_m)$ reflects the value of the outward current $\bar{I}$ as $t \to \infty$. The small parameter $\varepsilon(\overline{V}_m)$ defines the temporal properties of membrane outward currents and thus controls the duration of the AP and associated recovery processes. In the original Pushchino model with so-called standard parameters (see Chapter 5), the function $\varepsilon(\overline{V}_m)$ is presented as a two-part piece-wise function.

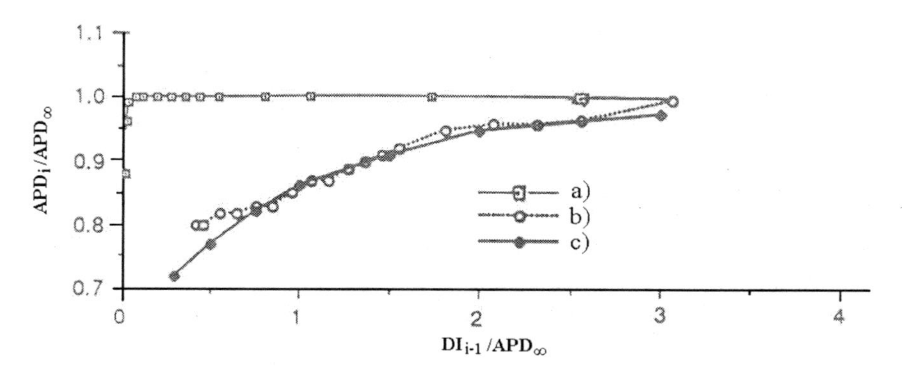

Fig. 4. APD restitution curves in dimensionless coordinates for (a) the FHN model with the standard set of parameters, (b) an actual cardiomyocyte (experimental results), and (c) the four-part piece-wise linear approximation of ε in the FHN model. APD_∞ denotes an AP duration occurring after a previous DI with sufficient length to allow all gating recovery processes to come to completion. The index i is the number of the excitation cycle.

The restitution curve obtained using this model in the course of computer simulation is shown in Fig. 4*a*. A key feature of note is that APDs are nearly constant except for small DIs. The restitution curve obtained from a real cardiomyocyte in a physiological preparation (Fig. 4*b*) shows a significant decrease of APD with shortening of the previous DI. In order to fit a desired restitution curve into the FHN model, a proposal [11] based on representing the function $\varepsilon(\overline{V}_m)$ as a four-part piece-wise linear approximation (see Chapter 5) was utilized:

$$\overline{\varepsilon} = \begin{cases} \varepsilon_1 & \text{if } \overline{V}_m < 0.01 \text{ and } \dfrac{d\overline{I}}{d\overline{t}} > 0 \quad \varepsilon_1 = 0.5 \\ \varepsilon_2 & \text{if } \overline{V}_m \geq 0.01 \text{ and } \dfrac{d\overline{I}}{d\overline{t}} > 0 \quad \varepsilon_2 = 0.02 \\ \varepsilon_3 & \text{if } \overline{I} > I_{\min} \text{ and } \dfrac{d\overline{I}}{d\overline{t}} \leq 0 \quad \varepsilon_3 = 0.5 \\ \varepsilon_4 & \text{otherwise} \qquad\qquad \varepsilon_4 = 0.018 \end{cases}.$$

The modified restitution curve for the FHN model with a four piece-wise linear approximation of the function $\varepsilon(\vec{V}_m)$ is shown in Fig. 4*c* and indicates that the proposed method gives a satisfactory result.

Prior to the presentation of 2D simulation results, a color and greycode legend for V_m (Fig. 5*a*) and I (Fig. 5*b*) is provided in order to facilitate understanding of quantitative values in the figures. These codes will be used through entire book.

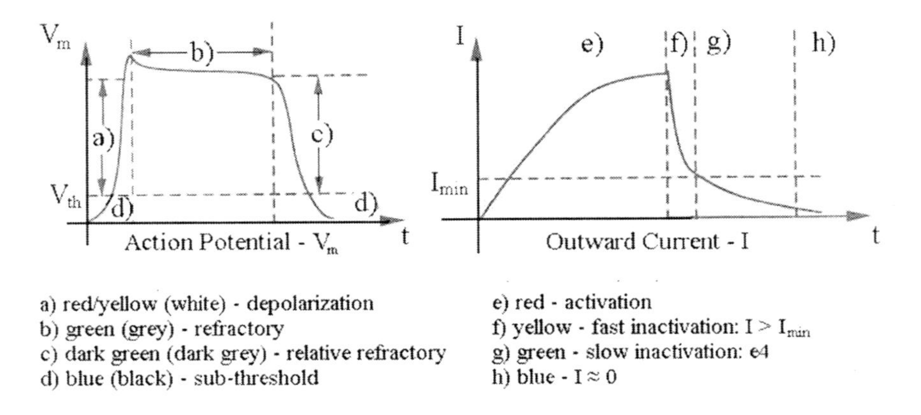

Fig. 5. Schematic representation of color and greycode levels (the latter in parentheses) designating the action potential (V_m) and outward current (I) phases.

Spiral wave initiation and subsequent propagation in a myocardium with a square configuration and composed of FHN AP models with Pushchino modifications [12] (with standard parameters) are shown in Fig. 6. In order to initiate the spiral wave, the S_1-S_2 protocol is utilized. This entails first providing S_1 stimulation to a group of nodes located in the left upper corner of the grid

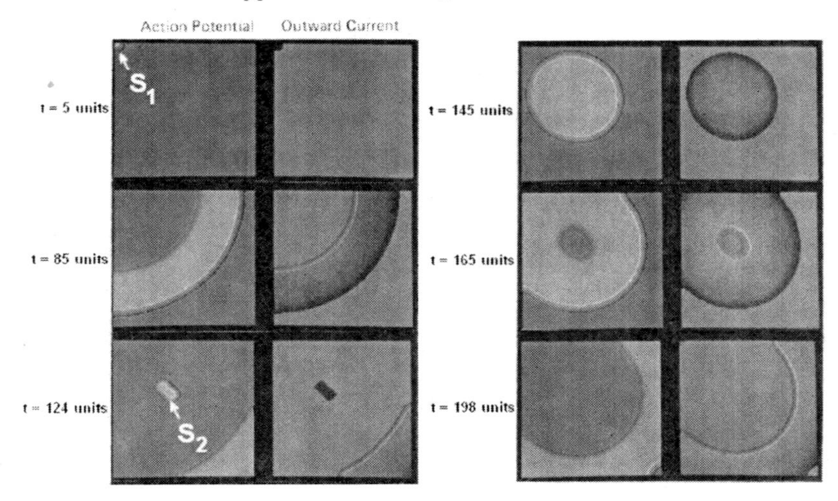

Fig. 6: Example of computer simulation results with a premature stimulus application to a 2D model of myocardium composed of FHN AP models with Pushchino modifications and standard parameters [12]. S_1 designates the stimulus initiating the original wave propagation and S_2 denotes the site of premature stimulation.

representing the myocardium, which leads to a semicircular wavefront arising after t = 25 time units. When t = 124 time units, the second (premature) stimulus, S_2, is applied as the wave begins to leave the myocardium from the corner opposite to the

S_1 site. As a result, a new circular wave appears (t = 145 time units) and continue to propagate until its radius reaches (t = 198 units) the grid borders and disappears. No spiral waves were observed in this circumstance. It was observed that a spiral wave can be initiated in these conditions only if the S_2 stimulus is applied to an area in close vicinity to the tail of the previous wave.

If the almost instantaneous inactivation of the outward current following the conclusion of the AP (which is present in the simulation shown in Fig. 6) is altered so that inactivation occurs more slowly (which is shown in Fig. 7), it then becomes possible to initiate spiral waves without changing the timing or location of the S_2 stimulus. This alteration is accomplished by introducing the four-part piece-wise linear approximation of the function $\varepsilon(V_m)$ [11] so that it fits more closely to the APD restitution curve of a real cardiomyocyte (see Figs. 4*a* and 4*b*).

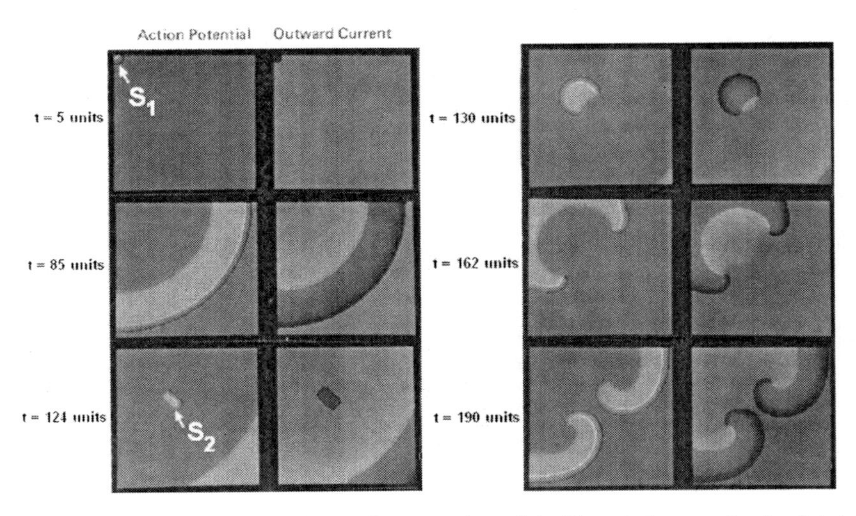

Fig. 7: Time course of the generation and propagation of double spiral waves in simulated 2D myocardium consisting of FHN AP models with the four piece-wise linear approximation of the $\varepsilon(V_m)$ function, which increases the inactivation time of the outward current under the effect of quinidine. Initial wave propagation is concentric. Premature beats are applied at the far edge of the central area of the window of vulnerability, which is shown in Fig. 8.

The window of vulnerability (WV) is defined as an area in the myocardium, generated in the wake of a previous propagating wave, where application of an appropriate premature stimulus can lead to the appearance of spiral waves. The shape of the WV follows from the morphology of the previous wave and its dimensions (see Fig. 8) depends on the restitution properties of membrane gate recovery processes. As these recovery processes become more prolonged, the WV will increase in size and facilitate spiral wave initiation. Moreover, recovery processes underlie both stationary and nonstationary (including breakup) wave propagation by generating stable or unstable functional heterogeneities, respectively, in myocardium that was initially uniform.

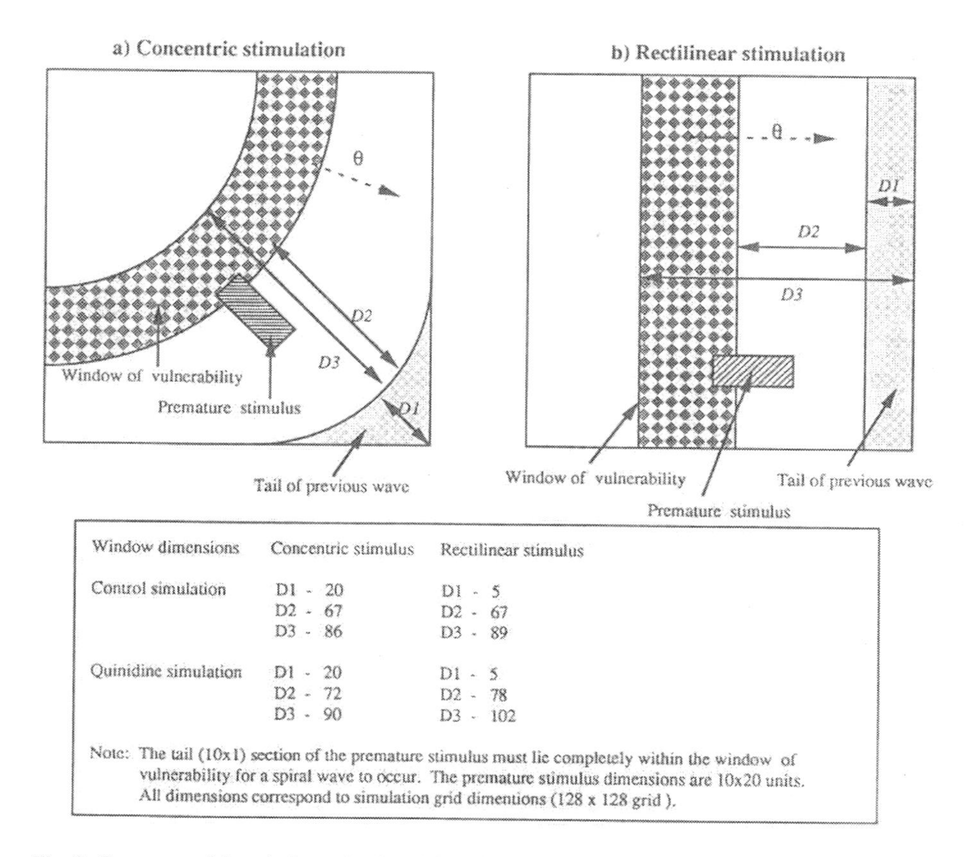

Window dimensions	Concentric stimulus	Rectilinear stimulus
Control simulation	D1 - 20 D2 - 67 D3 - 86	D1 - 5 D2 - 67 D3 - 89
Quinidine simulation	D1 - 20 D2 - 72 D3 - 90	D1 - 5 D2 - 78 D3 - 102

Note: The tail (10x1) section of the premature stimulus must lie completely within the window of vulnerability for a spiral wave to occur. The premature stimulus dimensions are 10x20 units. All dimensions correspond to simulation grid dimentions (128 x 128 grid).

Fig. 8: Geometry of the window of vulnerability following rectilinear and concentric stimulations.

Notable results were presented in [13] in relation to the initiation and study of stationary and nonstationary spiral waves in physiological experiments in relatively small (20 mm x 20mm x 0.5 mm) slices of sheep and dog epicardium. Spiral waves were first initiated via crossfield stimulation and were visualized spiral waves with a potentiometric dye in combination with charge coupled device (CCD) imaging technology. The majority of spiral waves observed in their experiments were anchored to small arteries or bands of connective tissue, which facilitate stationary propagation. In Fig. 9 (which is an excerpt from [13]), a sample of their physiological observations is provided with simulation results that utilized the FHN AP model with Pushchino modifications [12].

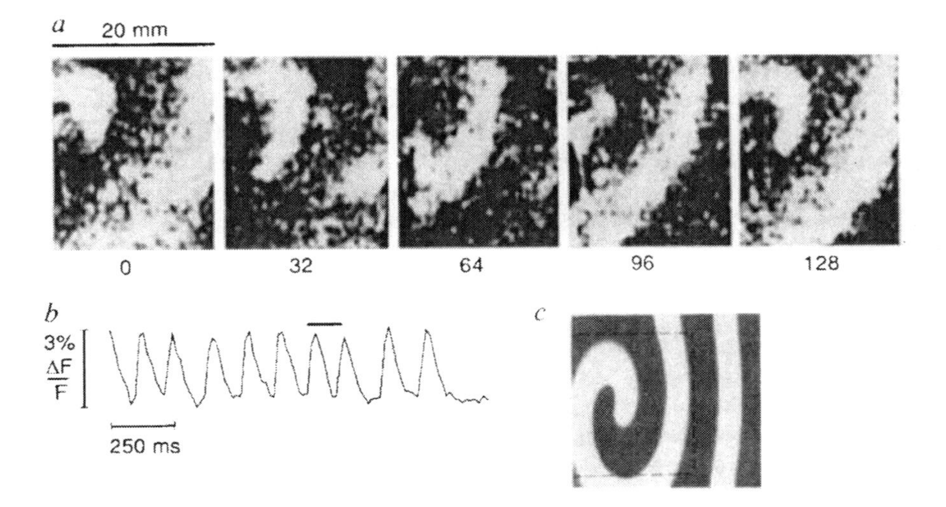

Fig. 9. (*a*) Clockwise-rotating spiral wave in canine epicardial muscle. White denotes maximal depolarization, black denotes resting potential, and the numbers represent the time in ms. (*b*) Time course of local activation on the upper left corner of the tissue during the last 10 cycles. Bar indicates time of recording in the (*c*) snapshot of a spiral wave obtained in a computer simulation using an AP model [12] in epicardium modeled as a grid of 96 x 96 nodes. The size of this grid is larger than the heart preparation shown in (*a*). The dashed line in (*c*) reflects the size of the preparation slices.

The analogy, even in a qualitative sense, between the physiological experiments and the computer simulation study is very superficial. The major reasons are threefold: (a) spiral wave rotation in the physiological experiment was observed to be anchored to anatomical obstacles, whereas rotation in the simulation study occurred in the center of unobstructed myocardium that only accounted for anisotropy; (b) the simplified AP model in the simulation study did not possess intracellular Ca dynamics and hence produced a shorter APD, especially during spiral wave propagation, in comparison to that measured in real epicardium, which experiences a potentiation of inward Ca-sensitive currents in response to intracellular Ca accumulation; and (c) the size of the experimental heart preparation is smaller than the corresponding grid utilized in the computer simulation, which itself increases border effects on stationary propagation. I mention these considerations, not to highlight the drawbacks of the presented results (which were highly reputable at the time they were published), but to warn readers that in order to claim two processes are analogous, it is first necessary to prove that they indeed satisfy certain criteria for analogy.

9.4.3. Simulations Using the Van Capelle and Durrer Model

The FHN AP model used in the aforementioned studies of wave propagation in a 2D model of myocardium is not an ionic AP model in the common sense of the term. Alternative simplified AP models (see Chapters 4 and 5) have been proposed that more closely resemble ionic AP models, but do not require a significant increase in computational resources. Among them is the Van Capelle and Durrer (VCD) simplified AP model [14], which has attracted wide attention [15, 16, 17]. In its original form, the VCD model was intended for implementation on a PC, leading to the imposition of certain restrictions on simulated AP and myocardium model parameters. These restrictions prohibit accurate reproduction of the major characteristics of the AP and hence create difficulties in the simulation of stationary spiral wave and fibrillation-like phenomena. Later in this section is a presentation of the results of computer simulations obtained with a VCD model that has modified parameters and an elimination of the restriction on grid sizes [15]. The latter accomplishment is achieved by replacing the PC with a parallel computer system (CM-2).

The VCD mathematical model of excitation-propagation in 2D myocardium with uniform anisotropy along the axes x and y incorporates two state variables: the membrane action potential V_m and the generalized excitability function $Y(t, V_m)$. The function Y varies between 0 (maximal excitability) and 1 (complete inexcitability). The following system of partial differential equations (PDEs) describes the behavior of these state variables in time and space:

$$C_m \frac{\partial V_m}{\partial t} = -(1-Y)\cdot f(V_m) - i_1(V_m) - i_{ext} \quad (14a)$$

$$T\frac{\partial Y}{\partial t} = Y_\infty(V_m) - Y \quad (14b)$$

$$f(V_m) = i_0(V_m) - i_1(V_m) \quad (14c)$$

$$i_{ext} = i_{st} + i_n \text{ and } i_n = \alpha_x\left(\frac{\partial^2 V_m}{\partial x^2} + A\frac{\partial^2 V_m}{\partial x^2}\right). \quad (14d)$$

In equations (14a)-(14d), C_m is the membrane capacitance, T is the time constant of the variable Y, $i_0(V_m)$ is the current-voltage relationship for a fully excitable membrane, $i_1(V_m)$ is the current-voltage relationship for a fully unexcitable membrane, $Y_\infty(V_m)$ is a steady state value of the function of Y, A is the anisotropy ratio (usually A = 1/4), i_{st} is the stimulus current, and i_n is the current from adjacent nodes in the grid model of the myocardium. The current i_n is also referred to as diffusion or local current. The parameters and major characteristics of the original and modified VCD models are presented in Table 1 and their associated APD restitution curves are presented in Fig. 10.

Table 1: Parameters and major characteristics of the VCD AP model and wave propagation in 2D myocardium: the original version of the model [14] and after its modification [15].

		Original Version	**After Modification**	**Comments**
Parameters	C_m (μF/cm^2)	10	1	
	T (ms)	0.5	As it was shown in [18]	Time constant of variable $Y(V_m)$
	A (unitless)	0.25	0.25	Anisotropy ratio: $A = \alpha_y / \alpha_x$
	α_x (kΩ^{-1})	0.00125-0.00133	Not changed	$\alpha_x = r/2R_i$ (see Chapter 5)
	n_x x n_y (unitless)	40 x 40	128 x 128	Grid size
	K (unitless)	1	4	Gain of $f(V_m)$ $K = f(V_m)_{mod} / f(V_m)_{orig}$
Characteristics	$\lvert V_m \rvert_{\max}$ (mV)	79.1	95.3	
	$\lvert \dot{V}_m \rvert_{\max}$ (V/s)	6.8	398	
	$APD_{DI \to \infty}$ (ms)	292	112	
	ρ (unitless)	25.3	467.7	AP relaxation coefficient $\rho = APD_\infty (V_m)_{\max} / V_{\max}$
	$Y(V_m, t)$	Solution of (14b)	Added logic function	$Y(V_m, t)_{mod} = 0$, $\partial V_m / \partial t > 0.01$. Otherwise, $Y(V_m, t)_{original}$
	$(\theta_{RL})_{long}$ (cm/s)	4.8	34.7	Post-modification data is presented according to [15]
	$(\theta_{RL})_{trans}$ (cm/s)	2.4	19	

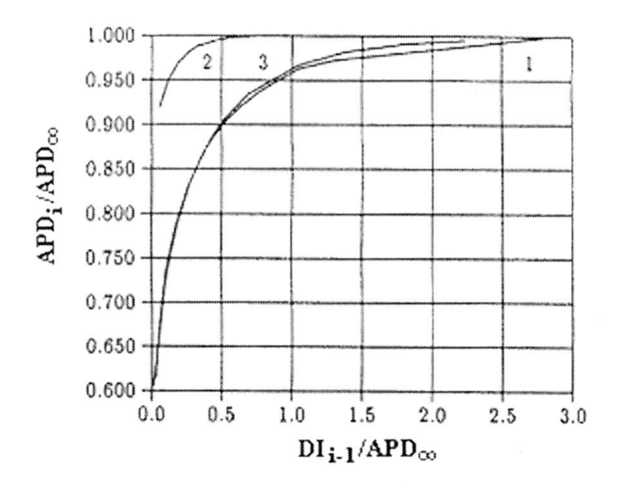

Fig. 10. The normalize APD restitution curves obtained from (1) the Luo-Rudy 1 AP model, (2) the original VCD AP model, and (3) the modified VCD AP model (3). $APD_i - APD_\infty$ is after an infinite diastolic interval and $APD_i - APD$ is after the diastolic interval DI_{i-1}.

Direct comparison of these data indicates that the original VCD model does not reproduce the majority of important physiological characteristics of the cardiac AP: gating recovery processes, the rate of depolarization, the relaxation coefficient value, and the rectilinear conduction velocity in both directions. Since the size of the myocardium is not sufficiently large to avoid border effects, a shortening of the WV occurs and causes difficulties in the initiation and support of stationary and nonstationary spiral wave propagation.

The aforementioned problems are rectified by modification of the characteristics and parameters of the VCD AP model (as shown in Table 1) and the fitting the APD restitution curves obtained from actual cardiomyocytes. These alterations facilitate the initiation and propagation of stationary and nonstationary spiral waves and in the presence of pacemaker activity or additional applied stimuli [19], spiral wave breakup is observed (see Fig. 11).

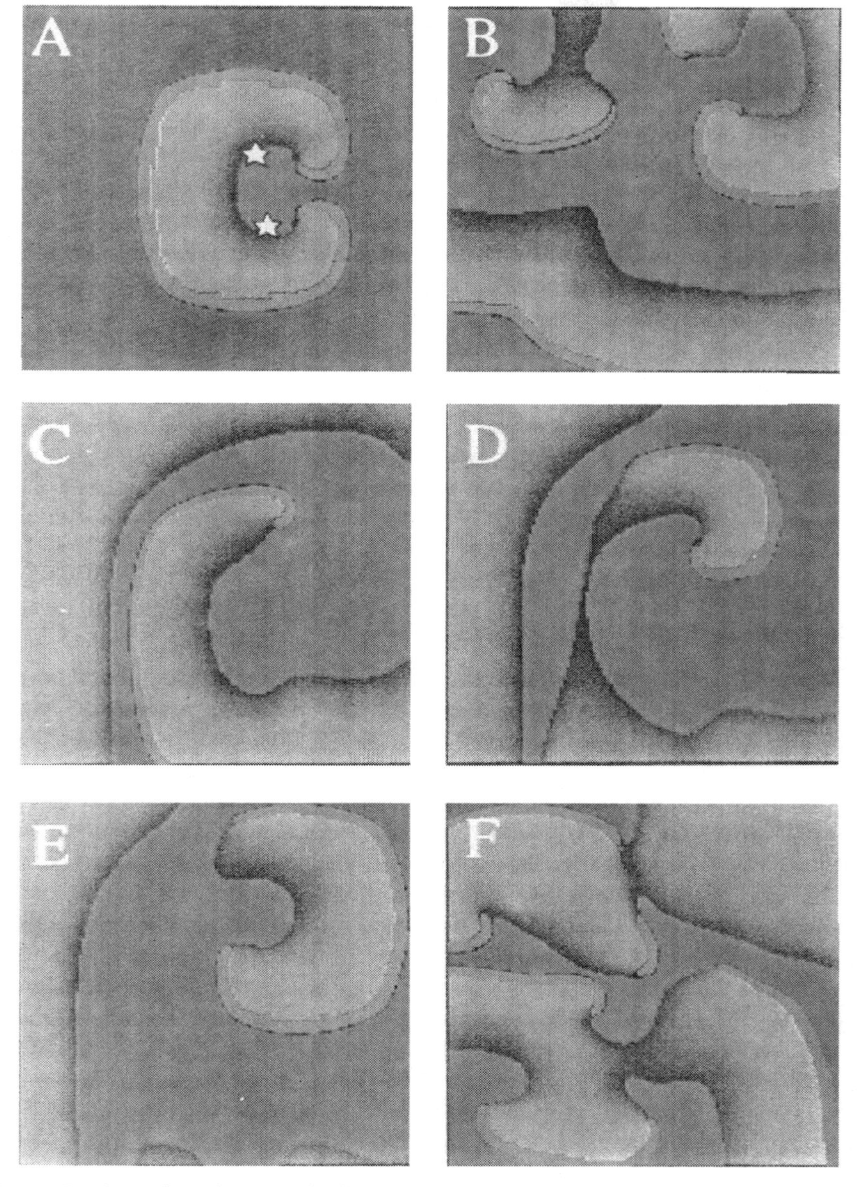

Fig. 11. Breakup of a spiral wave in 2D myocardium utilizing a modified VCD AP model [19]. (A) Generation of double spiral waves and application of two additional stimuli (stars). (B)Temporary breakup of the double spiral waves. (C) Reformation of a nonstationary spiral wave. (D)-(F) Progression to full breakup.

9.5. Propagation in 2D Myocardium with 2nd Generation AP Models

9.5.1. General Considerations

An entire spectrum of second generation AP models exists (see Chapter 4), but only two possess a complete representation of intracellular Ca dynamics (the Luo-Rudy 2, or LR2, model of the guinea pig ventricular AP and its modification by Chudin). As a result, these two AP models became the subject of a restricted set of comparative studies of spiral wave propagation in 2D isotropic myocardium and the associated findings are presented here. These simulations necessitated the computer implementation of the following PDE:

$$\frac{\partial V_m}{\partial t} = D\left(\frac{\partial^2 V_m}{\partial x^2} + \frac{\partial^2 V_m}{\partial y^2}\right) + I_{memb} + I_{st} \,. \tag{15}$$

In (15), all transmembrane currents, such as I_{memb} and I_{st}, are in units of μA/μF. The term $D = 0.001$ cm^2/ms represents the diffusion coefficient. In order to convert (15) into a closed form expression, it is necessary to include a system of ODEs which describes the behavior of all components of I_{memb} and the corresponding processes within intracellular compartments (see the appendix of chapter 4 for details regarding both AP models).

9.5.2. Simulations Using the LR2 AP Model

In order to make computer simulation feasible, the operator splitting algorithm, which facilitates the numerical solution of the PDEs on massively parallel supercomputers (see details in Chapter 6.4), was implemented on a CRAY-T3D [20]. This algorithm divides the integration of (15) into two portions: integration of the diffusion component of (15); and integration of the remainder of the (15) (the system of nonlinear ODEs), which can be performed at any point in space independently of its neighbors. The domain of integration was approximated by a 256×256 grid of nodes connected by coupling resistors. The grid is divided into strips along one axis and is subsequently divided among the parallel processors so that each processor is integrating (15) only in its own portion of the grid. Communication between processors is required only for the solution of the diffusion component of (15) and is implemented using the Message Passing Interface standard. An adaptive time step was utilized (varying between $\Delta t_{big} = 0.1$ ms and $\Delta t_{small} = 0.005$ ms) in order to integrate the system of ODEs during the overall interval $\Delta t = 0.1$ ms. The integration of the diffusion equation is performed twice during that interval with a time step $\Delta t_{diff} = \Delta t/2$ by using the explicit Euler method and a space step equal to $\Delta x = 0.025$ cm. The choice of the above parameters resulted in a rectilinear conduction velocity of 55 cm/s.

Spiral wave initiation entailed the application of a premature stimulus S_2 (in the form of a rectangular region) behind the tail of a propagating rectilinear wave (see Fig. 12 at time t_1). Since the premature excitation cannot propagate into the unrecovered region, a point q (where the wavefront is adjacent to the wavetail) appears.

As a result, the wavefront begins to circulate around this point and a spiral wave becomes established after a transient phase. The spiral wave morphology remains unaltered and the period of rotation changes slowly (from an initial value of 120 ms to 140 ms after 5 s). This phase lasts for approximately 6 s, after which the wave morphology begins to deform. The subsequent nonstationary circulation of the spiral wave continues for another 6 s and is followed by spiral wave breakup. All of these aforementioned processes are illustrated in Fig. 12.

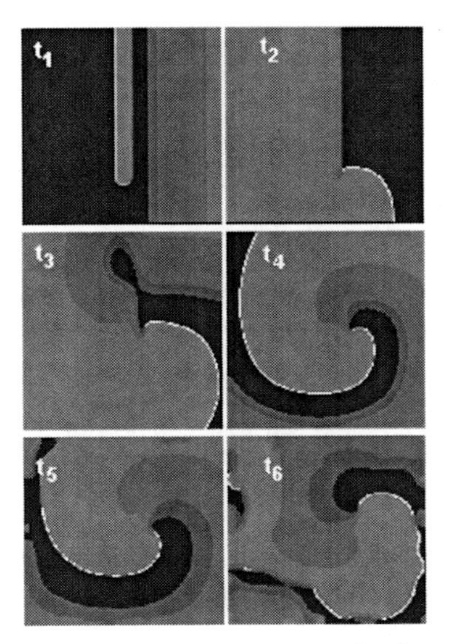

Fig. 12. Spiral wave propagation in the LR2 AP model. Application of a premature stimulus (t_1 = 250 ms). Spiral wave initiation (t_2 = 307 ms and t_3 = 507 ms) and the formation of a stationary spiral wave (t_4 = 1000 ms). Nonstationary propagation of the spiral wave due to intracellular Ca dynamics (t_5 = 11266 ms) and subsequent spiral wave breakup (t_6 = 15036 ms).

Results obtained from simulations using a solitary AP model suggested that the transition from the stationary to the nonstationary regime may be a consequence of gradual changes in complex intracellular Ca dynamics. Fig. 13 demonstrates that intracellular Ca dynamics, recorded at a single node in the tissue model during spiral wave circulation, is very similar to that observed in the solitary AP model simulations.

It was hypothesized that spiral wave deformation and breakup are caused by gradual changes in complex intracellular Ca dynamics during spiral wave rotation. The accumulation of $[Ca^{2+}]_i$ significantly increases the APD via amplification of the I_{NaCa} and others Ca-dependent currents. This APD prolongation leads to shortening of the diastolic interval and thus modifies wave propagation characteristics. The critical step in this process is abnormally large spontaneous Ca release from SR as a result of elevated Ca in the SR and myoplasm.

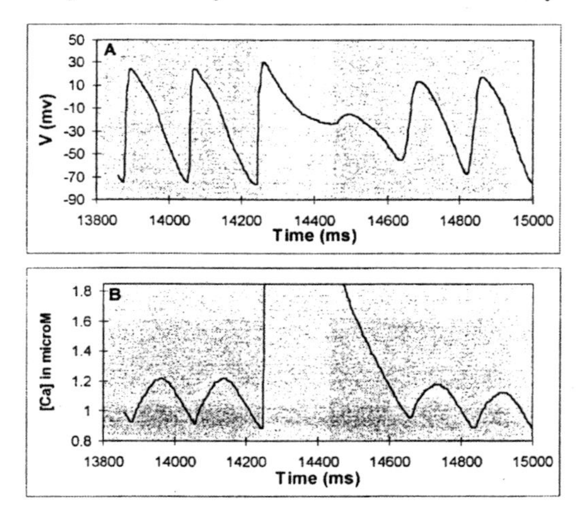

Fig. 13. Effect of Ca transients on APD during 2D wave propagation (A and B). Membrane potential and $[Ca^{2+}]_i$, respectively, recorded from a point with coordinates $(x, y) = (0.4$ cm, 4 cm), where the origin is located at the upper left corner. Values of $[Ca^{2+}]_i$ higher than 1.9 μM were indistinguishable in these 2D simulations, due to the employed coloring scheme (which explains why the graph presented in B is discontinuous).

To further investigate this hypothesis, simulations were conducted in which the L-type Ca channel was blocked. When the block was applied prior to the onset of wavefront breakup, a stationary spiral wave was obtained (see Fig. 14).

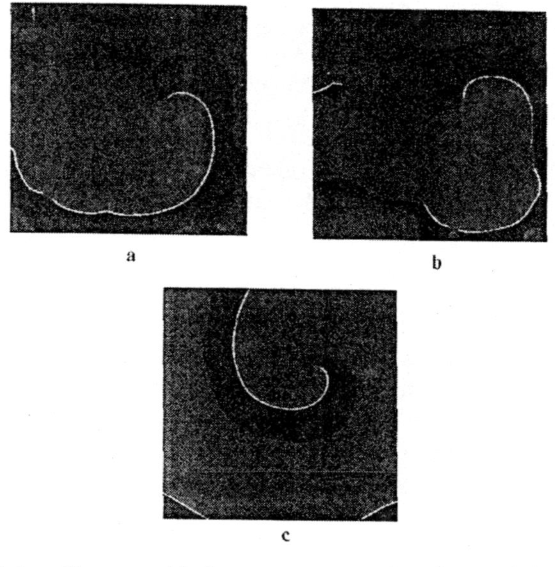

Fig. 14. Effect of L-type Ca current block on wave propagation characteristics. (a) Time = 13000 ms. (b) Wave front 1200 ms after (a) with no blockage of the L-type Ca current. (c) Wave front 1200 ms after (a) when the L-type Ca current block was applied. All 2D simulations were performed on a CRAY-T3D parallel supercomputer using the operator splitting algorithm, an adaptive time step, and a fixed space step equal to 0.025 cm.

However, if block was applied after wavefront breakup, multiple stationary spiral waves were obtained. It is possible to argue that the block of the L-type Ca channel not only affects intracellular Ca dynamics but also shortens the APD (by shortening the plateau phase of the AP) and thus stabilizes the spiral wave. In order to demonstrate that the change in intracellular Ca dynamics alone produced a stabilization of the spiral wave, simulations employing a clamped Ca distribution were performed. Although this distribution was spatially nonuniform, a stationary spiral wave was obtained and this strongly supported the Ca-dependent scenario of spiral wave breakup. In this respect, it is worthwhile to note that the APD restitution curve of the LR2 AP model (see [21]) has a slope less than unity for almost all diastolic intervals and is therefore unlikely to mediate spiral wave breakup.

In summary, intracellular Ca accumulation in the SR and myoplasm during high frequency stimulation conditions, such as spiral wave rotation, leads to irregular intracellular Ca dynamics when Ca in the overloaded SR is eventually released via spontaneous Ca release.

Studies of the LR2 model reveal that intracellular compartments become overloaded with Ca during conditions of rapid pacing. Spontaneous Ca release from the SR induces irregular changes in $[Ca^{2+}]_i$. As a consequence, APDs become abnormally prolonged due to the effect of the Na-Ca exchanger.

In a 2D grid of LR2 AP models, intracellular Ca accumulation during rapid spiral wave rotation causes spiral wave deformation and breakup. Wave breakup is preceded by the onset of highly irregular intracellular Ca dynamics. However, blockage of the L-type channel can restore stationary spiral wave propagation. The transition from nearly stationary to nonstationary spiral wave propagation and breakup is impossible to reproduce in 2D simulations with models that lack a detailed description of intracellular Ca dynamics.

9.5.3. Simulations Using the Chudin AP Model

AP propagation using the Chudin AP model [22] in a 2D isotropic uniform cardiac syncytium is described by the same PDE (15), with the corresponding initial and boundary conditions mentioned in the previous section. Indeed, the Chudin AP model reformulates only the intracellular Ca dynamics in the LR2 AP model (see the appendix of chapter 4 for a description of both models).

In order to make (15) closed, equations governing intracellular Ca dynamics and Hodgkin-Huxley-type gates are added. Computer simulation results illustrating spiral wave initiation and propagation in 2D myocardium with these modifications of intracellular Ca dynamics are shown in Fig. 15.

The diffusion coefficient D was chosen to provide a conduction velocity of ~55 cm/s for a solitary plane wave. As with the LR2 AP model, a premature S_2 stimulus was applied in order to obtain a reentrant spiral wave. Since the resultant excitation could not propagate into an unrecovered region, a point q (where the wavefront was adjacent to the wavetail) appeared. This led to the circulation of the wavefront around this point (see Fig. 15*A*), which formed the tip of the spiral wave (Fig. 15*B* and 15*C*).

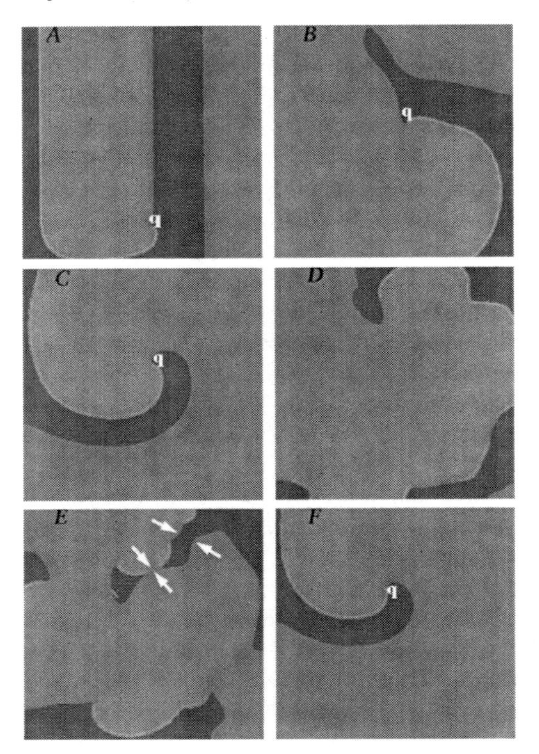

Fig. 15. Ca dynamics cause spiral wave breakup in a simulated 2D model of the myocardium. (A) Initiation of a spiral wave after application of a premature stimulus. (B) Appearance of transient reentry prior to the establishment of a stationary spiral wave. (C) Establishment of a spiral wave (t = 1275 ms). (D) Development of wavefront deformations (t = 4010 ms) leading to (E) breakup. (F) After spontaneous Ca release was blocked, multiple reentrant wavefronts coalesced back into a stationary spiral wave with a single wavefront.

The spiral wave rotated four times with a period of 170 ms and a diastolic interval of 20 ms, and then became nonstationary, with the wavefront progressively deteriorating for ~3000 ms (Fig. 15*D*). The breakup of the spiral wave was sensitive to the "gain" in the $[Ca^{2+}]_i$ sensitivity of various Ca-sensitive ionic currents. For example, increasing the $[Ca^{2+}]_i$ sensitivity of $I_{ns(Ca)}$ by decreasing $K_{m,ns(Ca)}$ from 1.2 μM to 0.9 μM facilitated breakup of the wavefront into a fibrillation-like state (Fig. 15*E*). During spiral wave rotation, each node located in the myocardium model is subjected to rapid excitation (CL = 170 ms), which is enough to cause intracellular Ca overload and spontaneous Ca release from SR (similar to the results observed in simulation studies of the solitary AP model).

Analysis of APs and Ca transients recorded from a local site in the tissue showed that the transition from the stationary to nonstationary regime started abruptly with an unusually large Ca transient (see Fig. 16*A*) due to spontaneous Ca release. This caused substantial prolongation of the APD (Fig. 16*A*, *top graph*), due to the increase in the inward components of I_{NaCa} and $I_{ns(Ca)}$, which led to marked shortening of the subsequent diastolic interval.

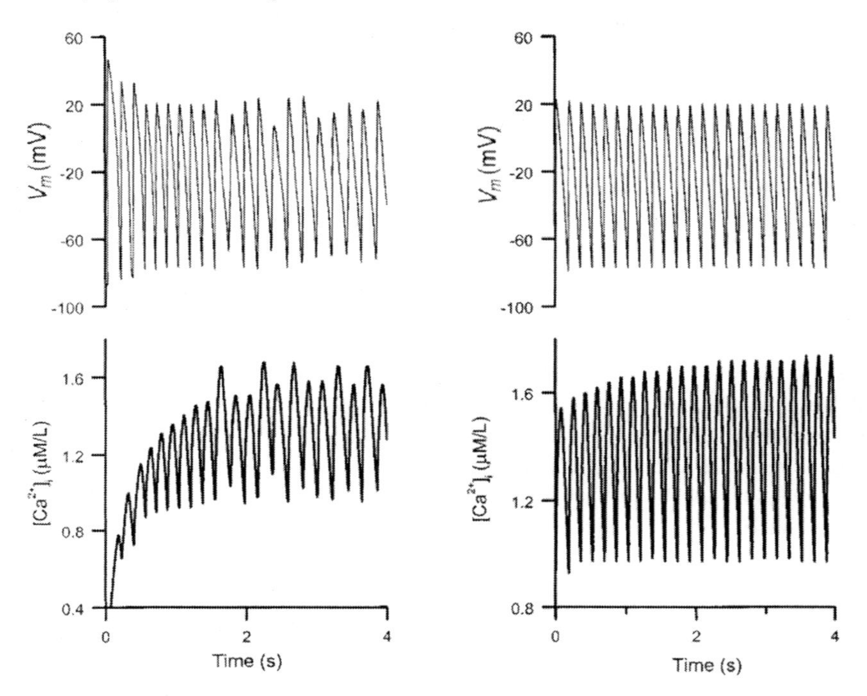

Fig. 16: Traces of transmembrane potential V_m (top), and $[Ca^{2+}]_i$ (bottom) measured at a site ($N_x = 100$, $N_y = 100$; origin at top left corner) in a simulated 2D myocardium. (A) V_m and $[Ca^{2+}]_i$ during the development of spiral wave breakup. (B) V_m and $[Ca^{2+}]_i$ after the block of spontaneous Ca release. Quasiperiodic oscillations of V_m disappeared when J_{spon} was blocked, as it is shown in Fig. 15*F*.

The short diastolic interval dramatically decreased the depolarization rate of the subsequent AP (by ~5-fold) due to the incomplete recovery of I_{Na} from inactivation, which slowed the conduction velocity of the wavefront. The short diastolic interval also shortened the subsequent APD due to its restitution properties, markedly altering the wavelength (the product of APD and conduction velocity) in this region. Conversely, the short diastolic interval and altered AP affected the intracellular Ca transient of the next beat, which further modified the AP via its feedback on Ca-sensitive currents.

If the interaction between $[Ca^{2+}]_i$ and the Ca-sensitive currents affecting the AP and conduction velocity is sufficiently strong, variations of restitution properties along the arm of the spiral wave grow to the point where the excitation wave can no longer propagate. Note particularly in Fig. 15*E* that wavebreaks occur at wavefront/wavetail interactions (see *arrows*). This causes the spiral wave to breakup, leading to a fibrillation-like state. In this scenario, spontaneous Ca release acts as a gain-enhancing mechanism between $[Ca^{2+}]_i$ and Ca-sensitive currents. As we mentioned above, if this gain was decreased by reducing the Ca sensitivity of Ca-sensitive currents (reducing the amplitudes of APD oscillations), then spiral wave breakup was prevented. Likewise, if the gain was decreased by eliminating spontaneous Ca release, spiral wave breakup also did not occur. In fact, if spiral wave breakup was allowed to develop with the spontaneous Ca release mechanism intact, its subsequent elimination caused the multiple reentrant wavefronts to coalesce back into a single stationary spiral wave (Figs. 15*F* and 16*B*). In this case, the tip of the spiral wave that was reformed exhibited a shift with respect to its original position (before start of nonstationary reentry), due to the redistribution of recovery processes during the nonstationary regime.

When the modified LR2 AP model was used to simulate 2D myocardium, it was found that intracellular Ca dynamics were directly responsible for causing the transition from stationary to violently meandering spiral wave reentry promoting wavebreak and a fibrillation-like state. This occurred because the complex temporal intracellular Ca dynamics resulted in spatial heterogeneities in $[Ca^{2+}]_i$, which amplified the inward components of I_{NaCa} and $I_{ns(Ca)}$ to produce, in turn, electrophysiological heterogeneities by prolonging the APD. These spatial regions of prolonged repolarization interact with the wavefront during the next rotation of the spiral wave, sharply decreasing conduction velocity and causing wavebreak. This is illustrated in Fig. 15, in which the red color represents the points on the wavefront where the absolute value of the I_{memb} current (see Eq. 15) is greater than 10 μA/μF (the approximate threshold corresponding to significant participation of I_{Na} in wavefront propagation). Breaks in the red line indicate slow propagation supported by the L-type Ca current (where I_{Na} is highly inactivated) or conduction failure.

The qualitative nature of these results remain robust with respect to various aspects of CICR current such as expressions for $P(V_m)$, γ, and the value of G_{rel}. Moreover, despite the different formulation of intracellular Ca dynamics and different morphology of Ca transient, the original LR2 AP model gave qualitatively similar results; that is, when its Ca dynamics were operational, spiral wave breakup occurred due to the Ca instability [20].

These findings are generally consistent with studies implicating cardiac restitution properties as key determinants of spiral wave instability and breakup. The effects of intracellular Ca may operate dynamically by promoting functional electrophysiological heterogeneities. By modulating various Ca-sensitive currents, intracellular Ca levels locally alter cardiac restitution properties. If the "gain" between intracellular Ca and Ca-sensitive currents affecting restitution is sufficiently high, intracellular Ca dynamics may promote instability. Moreover, it was shown that intracellular Ca dynamics under conditions of stationary spiral wave propagation may cause such phenomena as Ca and APD alternans [7] and the appearance of single point and clusters of EADs [8,9] and DADs [10] in tissue, which may facilitate spiral wave instability.

9.6. Appendix: Derivation of Curvature for a Spiral Wave

The derivation for the Natural Equations provided below is based on the expressions in (1) for θ and v and the following Frenet formulas [23]: $\dfrac{\partial \overline{\tau}}{\partial s} = K\overline{n}$ and

$\frac{\partial \overline{n}}{\partial s} = -K\overline{\tau}$. After performing the mathematical transformations described by Zykov [1], the Natural Equations presented in (5) and (6) are obtained:

$$\frac{d\theta}{ds} + K(s)v(s) = \omega$$

$$\frac{dv}{ds} - K(s)\theta(s) = 0,$$

with the boundary conditions $\theta(s)\big|_{s=0} = \theta_{CR}$ and $v(s)\big|_{s=0} = 0$. In this circumstance, $s = 0$ at the point Q

The curvature of a stationary wavefront can be determined using the Natural Equations. This is possible by multiplying (5) by $\theta(s)$, which yields

$$\theta(s)\frac{d\theta(s)}{ds} + \theta(s)K(s)v(s) = \theta(s)\omega. \tag{16}$$

A subsequent substitution of (6) into (16) produces the transformed expression $\theta(s)\frac{d\theta}{ds} + v(s)\frac{dv}{ds} = \theta(s)\omega$. By inspection, it is evident that the left-hand side of the transformed equation is equivalent to $\frac{1}{2}\frac{\partial}{\partial s}\left[\theta^2(s) + v^2(s)\right]$. Therefore, this equation can be rewritten as

$$\frac{d}{ds}\left[\theta^2(s) + v^2(s)\right] = 2\theta(s)\omega. \tag{17}$$

Since it follows from (1) that θ, v, and u are all associated by the relation

$$\theta^2(s) + v^2(s) = u^2, \tag{18}$$

the integration of (17) with respect to s and within the limits 0 to S gives $\int_0^s \frac{d\left(u^2(s)\right)}{ds} ds = 2\omega \int_0^s \theta(\zeta)d\zeta$. Evaluation of the left-hand side of this integral expression produces the following equation:

$$u^2(s) = 2\omega \int_0^s \theta(\zeta)d\zeta + u^2(0). \tag{19}$$

A substitution of (18) into (19) permits an integral expression for $v^2(s)$: from (17) and (19) as:

$$v^2(s) = 2\omega \int_0^s \theta(\zeta)d\zeta + u^2(0) - \theta^2(s). \tag{20}$$

Rearrangement of (5) and appropriate substitutions of the transformed expression of (16) and (20) yields an explicit expression for $K^2(s)$, the curvature of the spiral wave:

$$K^2(s) = \frac{\left(\omega - \frac{d\theta}{ds}\right)^2}{2\omega \int_0^s \theta(\zeta)d\zeta + u^2(0) - \theta^2(s)}. \tag{21}$$

Equation (21), together with the considerations of kinematic theory presented in the beginning of this chapter, allows qualitative estimation of the overall spiral wavefront morphology. Fig. 17, which is nearly unchanged from [1], illustrates the dependence of the key determinants (θ, v, K) of spiral wave morphology on s. It is necessary to note that the point S=q (the origin of the plots) represents a distinct point where the wavefront and wavetail of the stationary spiral wave are joined.

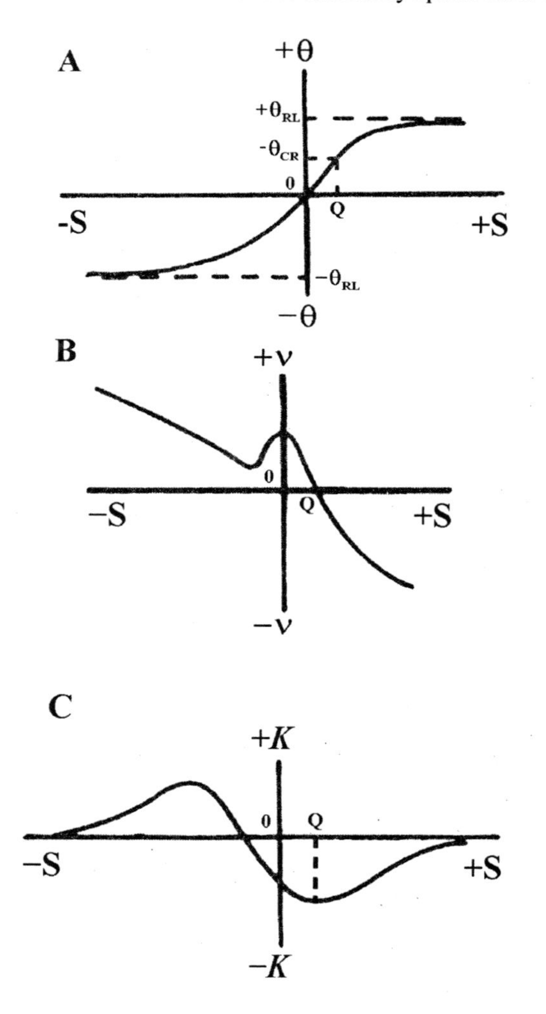

Fig. 17. Parameters that characterize the morphology of a stationary spiral wave Distribution of θ, the normal component of the conduction velocity u. (B) Distribution of v, the tangential component of u. (C) Distribution of K, the wavefront curvature.

9.7. References

1. Zykov, V.S., *Simulation of wave process in excitable media*. Nonlinear science: theory and applications, ed. A.V. Holden. 1987, Manchester and New York: Manchester University Press.
2. Kogan, B.Y., W.J. Karplus, B.S. Billet, and W. Stevenson, *Excitation wave propagation within narrow pathways: geometric configurations facilitating unidirectional block and reentry*. Physica D, 1992. **59**: 275-296.
3. Courtemanche, M. and A.T. Winfree, *Re-entrant rotating waves in a Beeler-Reuter based model of two- dimensional cardiac conduction*. Int J Bifurc Chaos, 1991. **1**: 431-444.
4. Panfilov, A.V. and A.V. Holden, *Spatiotemporal irregularity in a two-dimensional model of cardiac tissue*. Int J Bifurc Chaos, 1991. **1**: 219-225.
5. Qu, Z., J.N. Weiss, and A. Garfinkel, *Cardiac electrical restitution properties and stability of reentrant spiral waves: a simulation study*. Am J Physiol Heart Circ Physiol, 1999. **45**: H269-H283.
6. Courtemanche, M., L. Glass, and J.P. Keener, *Instabilities of a propagating pulse in a ring of excitable media*. Phys Rev Lett, 1993. **70**: 2182-2185.
7. Shiferaw, Y., M.A. Watanabe, A. Garfinkel, J.N. Weiss, and A. Karma, *Model of intracellular calcium cycling in ventricular myocytes*. Biophys J, 2003. **85**: 3666-3686.
8. Kogan, B., S.T. Lamp, and J.N. Weiss, *Role of intracellular Ca dynamics in supporting spiral wave propagation*, in *Modeling and Simulation*, G. Bekey and B. Kogan, Editors. 2003, Kluwer Academic Publishers: Norwell, MA. p. 177-193.
9. Huffaker, R.B., J.N. Weiss, and B. Kogan, *Effects of early afterdepolarizations on reentry in cardiac tissue: a simulation study*. Am J Physiol Heart Circ Physiol, 2007. **292**: H3089-H3102.
10. Samade, R., *Personal Communication*. 2008.
11. Kogan, B.Y., W.J. Karplus, B.S. Billet, A.T. Pang, H.S. Karagueuzian, and S.S. Khan, *The simplified Fitzhugh-Nagumo model with action potential duration restitution: effects on 2D-wave propagation*. Physica D, 1991. **50**: 327-340.
12. Pertsov, A.M., E.A. Ermakova, and A.V. Panfilov, *Rotating spiral waves in a modified Fitz-Hugh-Nagumo model*. Physica D, 1984. **14**: 117-124.
13. Davidenko, J.M., A.V. Pertsov, J.R. Salomonsz, W. Baxter, and J. Jalife, *Stationary and drifting spiral waves of excitation in isolated cardiac muscle*. Nature, 1992. **355**: 349-351.
14. van Capelle, F.J.L. and D. Durrer, *Computer simulation of arrhythmias in a network of coupled excitable elements*. Circ Res, 1980. **47**: 454-466.
15. Kogan, B.Y., W.J. Karplus, and M.G. Karpoukhin. *The Van Capelle and Durrer model of cardiac action potential generation and 2D propagation: modification and application to spiral wave propagation*. in *Proceedings of the Society of Computer Simulation*. 1996. San Diego, CA. p. 106-112.
16. van Capelle, F.J.L., *Propagation and reentry in two dimensions*, in *Cardiac Electrophysiology: From Cell to Bedside*, D.P. Zipes and J. Jalife, Editors. 1990, WB Sauders Co: Philadelphia, PA. p. 175-182.
17. Landau, M., P. Lorente, J. Henry, and S. Canu, *Hysteresis phenomena between periodic and stationary solutions in a model of pacemaker and nonpacemaker coupled cardiac cells*. J Math Biol, 1987. **25**: 491-509.
18. Panfilov, A.V. and A.V. Holden, *Computer simulation of re-entry sources in myocardium in two and three dimensions*. J Theor Biol, 1993. **161**: 271-285.
19. Garfinkel, A., P.-S. Chen, D.O. Walter, H.S. Karagueuzian, B. Kogan, S.J. Evans, M. Karpoukhin, C. Hwang, T. Uchida, M. Gotoh, O. Nwasokwa, P. Sager, and J.N. Weiss, *Quasiperiodicity and chaos in cardiac fibrillation*. J Clin Invest, 1997. **99**: 305-314.
20. Chudin, E., A. Garfinkel, J. Weiss, W. Karplus, and B. Kogan, *Wave propagation in cardiac tissue and effects of intracellular calcium dynamics (computer simulation study)*. Prog Biophys Mol Biol, 1998. **69**: 225-236.
21. Zeng, J., K.R. Laurita, D.S. Rosenbaum, and Y. Rudy, *Two components of the delayed rectifier K+ current in ventricular myocytes of the guinea pig type. Theoretical formulation and their role in repolarization*. Circ Res, 1995. **77**: 140-152.

22. Chudin, E., J. Goldhaber, A. Garfinkel, J. Weiss, and B. Kogan, *Intracellular Ca(2+) dynamics and the stability of ventricular tachycardia.* Biophys J, 1999. **77**: 2930-2941.
23. Frenet, F., *Sur les corbes a double courbure.* J Math Pur Appl, 1852. **17**: 437-447.

Chapter 10. Excitation Wave Propagation in Narrow Passes

10.1. Introduction

Direct physiological evidence [1-4] exists that occurrences of arrhythmia are commonplace in the presence of infarct scars, where regions of normal and excitable myocardium are interspersed with regions of unexcitable myocardium. These regions form narrow and wide pathways for wave propagation and each of these pathways assumes a configuration that can be categorized into a particular type of border geometry.

The concept of critical curvature of the wavefront (introduced in chapter 9) provides a connection between pathway border geometry and the properties of surviving myocardium within the pathway and the appearance of a unidirectional conduction block. The conduction block facilitates the appearance of reentrant arrhythmias, which can in turn, lead to ventricular fibrillation.

This concept proves valid, at least qualitatively, for any type of AP and myocardium. Here, this discussion is restricted to wave propagation through narrow pathways with three idealized geometric configurations (parallel borders, tapered borders and combinations of parallel and tapered borders) and three types of boundary conditions (impermeable, fully unexcitable, and ones with decreased excitability). Special consideration is given for different sets of conditions that facilitate the appearance of a unidirectional block and propagation reentry through these pathways. The roles of anisotropy in normal myocardium and AP recovery processes in reentry formation are illuminated.

The computer simulation results presented in the following sections were obtained using a modified FHN simplified AP model in a model of 2D myocardium, which was represented as a grid of 128×128 nodes, that was solved numerically on a massively parallel supercomputer (the CM-2 from the Thinking Machine Corporation).

10.2. Theoretical Considerations

Theoretical considerations are not dependent on the properties of a particular model: their results have generalized significance and they illustrate the qualitative characteristics of processes related to wave propagation, within the context of the utilized assumptions. Computer simulations are employed in order to obtain the quantitative characteristics of a particular choice of both the AP and myocardium models, as well as to verify the assumptions used.

Three idealized types of border geometry in narrow pathways are encompassed by theoretical considerations first presented by Kogan *et al.* [5]: *parallel borders*, *tapered borders*, and *combinations of parallel and tapered borders*. For each geometrical configuration, we consider the following three boundary conditions:

B.Ja. Kogan, *Introduction to Computational Cardiology: Mathematical Modeling and Computer Simulation*, DOI 10.1007/978-0-387-76686-7_10,

1. $\frac{\partial V_m}{\partial n} = 0$ Impermeable border zone (connective tissue)

2. $\frac{\partial V_m}{\partial n} = grad\ V_m$ Fully unexcitable border zone (electrolytes present with scarred myocardium)

3. $0 < \frac{\partial V_m}{\partial n} < grad\ V_m$ Decreased excitability in the border zone (scars with a mixture of viable and dead cardiomyocytes)

At an arbitrary point, s, on the wavefront, an expression given by $\alpha = f(s)$ can be formulated, where α is the angle between the tangent at a particular point s and the x axis component of the V_m gradient. These relationships are illustrated in Fig. 1.

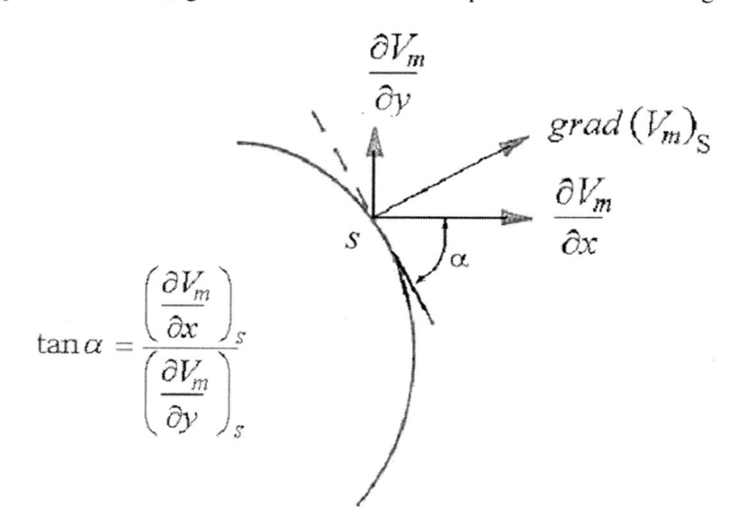

Fig. 1. Determination of the wavefront morphology by using components of the gradient of V_m.

The gradient components at each point s on the wavefront are determined via calculation of the Laplacian during simulations, where they are computed as an intermediate result. It follows from the expression for tan (α) that when the components of grad (V_m) separately tend to zero for some point s, the angle of the tangent at this point tends either to zero or 2π (when $\partial V_m/\partial x = 0$) or to $\pm\pi/2$ (when $\partial V_m/\partial y = 0$). Here, it is assumed that the dimensions of these narrow paths in myocardium are considerably larger than those of a cardiomyocyte. This limits the parameters of the study to macro processes.

The three types of boundary conditions mentioned above may be applied to various types of excitable media; in the particular case of myocardium, they reflect three possible situations. First, the current between myocardium in the excitable zone and the border zone, within the narrow pathway, may be zero, analogous to the situation when the coupling resistors have infinite resistance (which is possible when connective tissue is present). This situation is the impermeable, or zero-flux, boundary condition. A second situation occurs when current does flow to the border zone from the excitable zone, but fails to elicit a response since the border zone myocardium is fully unexcitable there (the expression $\partial V_m / \partial n$ = grad (V_m) represents the boundary condition). This can occur if the border zones consist of an electrolyte solution, which decreases the coupling resistance between the excitable and border zones. The third situation is characterized by a border zone consisting of myocardium with markedly depressed excitability or a significant prolongation of the time necessary to recover from inactivation. Cardiomyocytes exhibiting the latter characteristics have been observed to coexist in the same infarct zone as cardiomyocytes with relatively normal membrane characteristics [1, 5, 6, 7]. This type of boundary condition represents decreased excitability of the border zone, which is described the inequality $0 < \partial V_m / \partial n <$ grad (V_m).

Normal ventricular myocardium tissue is anisotropic, due to greater intercellular resistance along the transverse axis of cardiomyocytes in comparison to the longitudinal axis. When infarct scars are present, anisotropic wave propagation is unaltered and can be accentuated [8]. Uniform anisotropy is modeled by appropriately altering resistances between excitable elements (nodes) in the transverse direction.

10.3. Propagation Inside Narrow Pathways

In this section, we discuss three idealized geometries of narrow pathways:

1. parallel borders
2. tapered borders
3. combinations of parallel and tapered borders.

Three types of boundary conditions that arise from the properties of the border zone are considered for each geometry:

1. impermeable
2. unexcitable
3. with decreased excitability.

10.3.1. Propagation Inside Narrow Pathways with Parallel Borders

In a pathway with parallel borders and impermeable boundary conditions, the component of grad (V_m) along the y-axis ($\partial V_m / \partial y$) is zero and the angle of the tangent to the wavefront at all points of the border is $\pi / 2$. Assuming uniform isotropy inside the pathway, the wavefront that propagates within it must be rectilinear (see Fig. 2a) with a curvature $K = 0$.

For the boundary conditions $\partial V_m / \partial n$ = grad (V_m), it may be assumed that the points along the border are held at rest potential; therefore, the grad (V_m) component $\partial V_m / \partial x$ and the angle α are equal to zero at all points along the border. Inside the narrow pathway, the $\partial V_m / \partial x$ component of grad (V_m) and the angle α increase until their maximum values (grad (V_m) = [grad (V_m)]$_{\max}$ and $\alpha = \pi / 2$) are achieved at the midpoint of the wavefront. It is therefore reasonable to approximate the wavefront inside a narrow path as a semi-circle with a radius equal to $W / 2$ and a curvature $K = 2 / W$.

When the boundary conditions are $0 < \partial V_m / \partial n <$ grad (V_m), the angle of the tangent at the border points is determined by the ratio of the corresponding components of grad (V_m). Since the pathway is narrow, the wavefront can be approximated as a portion of a circle with a radius $R_b = (W / 2) \cdot \cos(\alpha_b)$ and a curvature given by

$$K_b = \frac{2}{W}\cos(\alpha_b) = \frac{2}{W}\frac{1}{\sqrt{1+\tan^2(\alpha_b)}}, \quad (1)$$

where

$$\tan\alpha_b = (\partial V_m / \partial x)_b / (\partial V_m / \partial y)_b . \quad (2)$$

Similarly, for any point i in Fig. 2 (except point 0 on the wavefront) the curvature is found to be

$$K_i = \frac{2}{W}\cos(\alpha_i) = \frac{2}{W}\frac{1}{\sqrt{1+\tan^2(\alpha_i)}}, \quad (3)$$

where

$$\tan(\alpha_i) = (\partial V_m / \partial x)_i / (\partial V_m / \partial y)_i . \quad (4)$$

All geometric constructions that are necessary to derive these formulas are shown in Fig. 2a, 2b, and 2c.

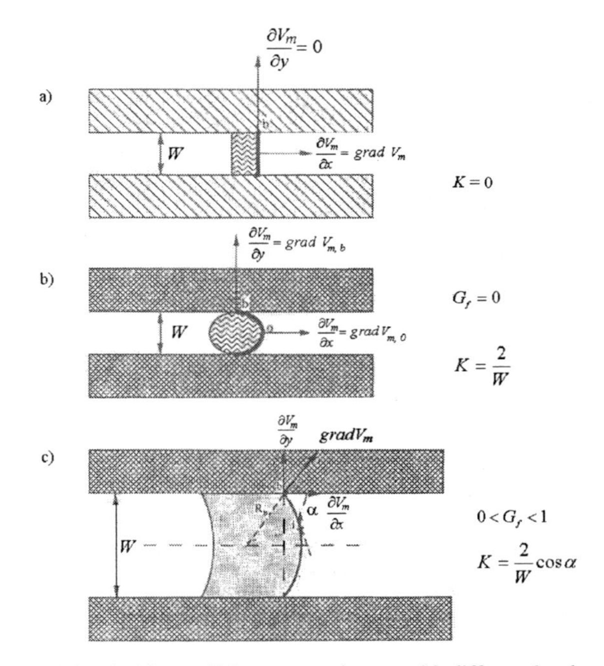

Fig. 2. Wave propagation inside parallel narrow pathways with different border zone conditions: (a) impermeable, (b) fully unexcitable, and (c) decreased excitability.

10.3.2. Propagation Inside Narrow Pathways with Combinations of Parallel and Tapered Borders

Simulated geometric configurations of narrow pathways possessing both parallel and tapered borders are shown schematically in Fig. 3. By changing the value of the angle β from zero to $\pi / 2$, it is possible to create a continuous transition from a narrow pathway with parallel borders to one with a combination of parallel and tapered borders. In this case $\beta = \pi / 2$ corresponds to the abrupt opening of the narrow pathway to the unrestricted right half-plane of excitable myocardium. Theoretical considerations and computer simulations with the FHN model show that in a pathway with parallel borders, stationary waves propagate with a rectilinear front.

The wavefront conduction velocity is constant regardless of channel width. Since the propagating wavefront must be perpendicular to the borders at all points, the wavefront at the points *a-a* of the expanding borders can be considered (assuming that the excitable myocardium is isotropic everywhere), as a first

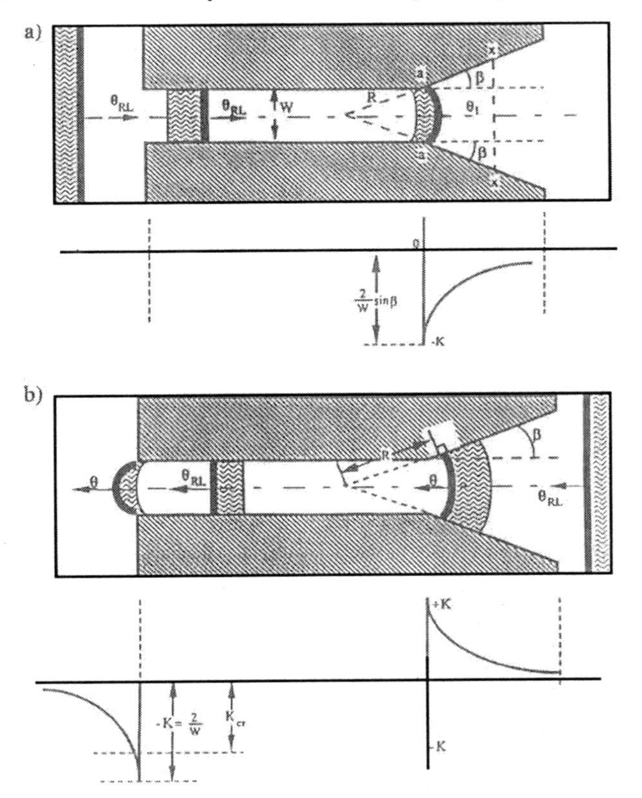

Fig. 3. Propagation of wavefronts in narrow pathways with a combination of parallel and tapered border geometries and an impermeable border. (a) Propagation from a pathway with parallel borders to a tapered one. (b) In the same configuration as (a), but with propagation in the opposite direction.

approximation, to be a circular arc. The simple geometric drawing shown in Fig. 3a, gives

$$R = W/(2\sin(\beta)) \tag{5}$$

and

$$K = 1/R = (2\sin(\beta))/W\,, \tag{6}$$

where R is the radius of the wavefront at the pathway opening, W is the width of the pathway with parallel borders, β is the angle of border inclination, and K is the curvature of the wavefront at the points *a-a*.

Eq. (6) specifies a family of sinusoidal curves at an amplitude of $A = 2 / W$ and is valid for all points *x-x* on the border of a tapered opening. It is only necessary to substitute in (6) the value of $W / 2$ by its corresponding value $W_{x\text{-}x} / 2$ at the points *x-x* (see Fig. 3a). The geometry of this figure gives

$$W_{x-x}/2 = (W/2) + (L - L_x)\tan(\beta), \tag{7}$$

where L is the full length of the tapered portion of the pathway and L_x is the distance from the wide end to the points "*x-x*" in the tapered portion. Thus, the curvature K_x will be

$$K_x = \pm\frac{\sin(\beta)}{(W/2) + (L - L_x)\tan(\beta)}. \tag{8}$$

The signs + and - correspond to concave and convex morphologies in the propagating wavefronts, respectively.

As discussed above, a critical value of the wavefront curvature exists such that propagation becomes impossible above this value (i.e. there is a conduction block). Physically, this represents that condition when the excitable cardiomyocytes (sources) fail to transmit an electrical current sufficient to depolarize neighboring cardiomyocytes (sinks) to the threshold of excitation. This critical value depends on the selected AP model and its active and passive parameter values. Let us assume that the value of K_{cr} is known; thus, we obtain from (6)

$$K_{cr} = \frac{2}{W}\sin(\beta) \tag{9}$$

and

$$\beta = \arcsin\left(\frac{K_{cr}W}{2}\right). \tag{10}$$

When K_{cr} is constant, (10) gives the relation of β_{cr} to W_{cr}. Fig. 4 shows the results of computer simulations used to determine this dependency for the FHN model with isotropic media. The results in Fig. 4 were used to test the assumption that when the dimensions of a channel are close to critical, the wavefront in a tapered channel can be approximated by a circular arc.

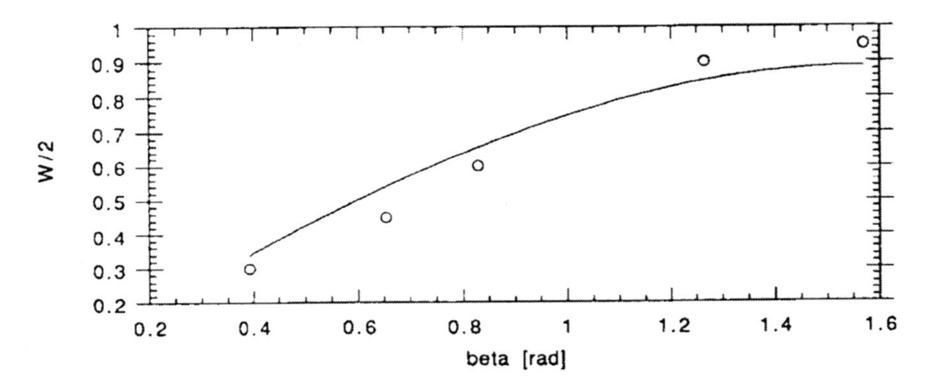

Fig. 4: Critical angle β for propagation block at narrow pathways exits for various pathway widths in isotropic media. A nonlinear curve fit to a sine curve, given by $W / 2 = 0.89 \sin(\beta)$, is shown for the data and produces a regression correlation $R^2 = 0.939$.

A non-linear regression fit of this computer simulation data to the predicted sine function gives a sufficiently accurate regression correlation of $R^2 = 0.939$.

Fig. 5 shows the same phenomena for a wide range of anisotropy ratios. As the anisotropy ratio (D_x / D_y) increases, the critical width of the pathway increases as well. This phenomenon can be explained by a decrease in the current consumption of the neighboring cardiomyocytes, located in the transverse direction, when tissue anisotropy is increased. This leads to a decrease of wavefront curvature at the exit of the narrow pathway, as if the effective width of the pathway opening has decreased.

Direct measurement of the critical wave curvature is very difficult in physiological experiments. Determination of the critical curvature via computer simulations

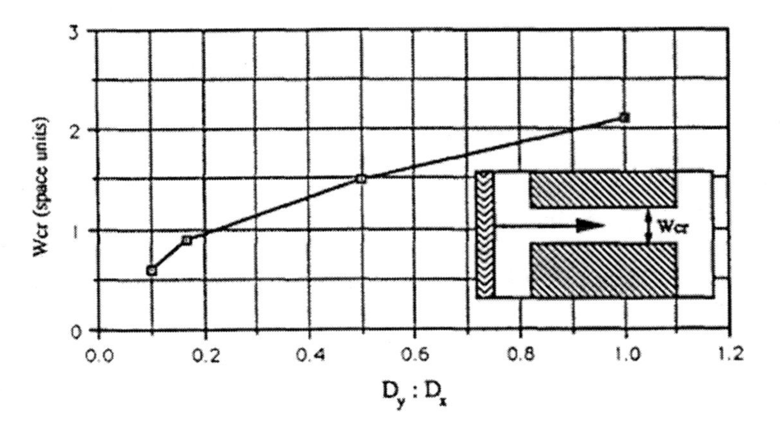

Fig. 5. Relationship of critical width (β is held constant at 90°) to the anisotropic ratio, when conduction block occurs in narrow paths with impermeable borders.

requires a comparatively large amount of calculations, which increases with the complexity of the model. Equation (6) permits a measurement of the width of a narrow pathway instead of the wavefront curvature in order to obtain the critical value of wavefront curvature. In isotropic media inside narrow pathways and $\beta = \pi / 2$, the curvature is given by $K = 2 / W$. This formula does not account for variation of the wavefront conduction velocity in the original rectilinear wave when it is entering and exiting the narrow pathway; resultant effects are addressed below.

Methodical alterations of the narrow pathway width during a computer simulation allows for the determination of the value $W = W_{cr}$, at which propagation through the opening becomes impossible. The values for K_{cr} obtained by Zykov [9] for the FHN model, using an approximate formula and an iterative solution of the original equation ($K_{cr} = -0.79$) and obtained in the above approach ($K_{cr} = -0.83$), are in close agreement.

Let us return to the configuration shown in Fig. 3a, where W is chosen so that $W = W_{cr}$ and $\beta = \pi / 2$ (left entrance to the narrow path). When the wave propagates from left to right, its curvature remains equal to zero throughout the narrow pathway until it reaches the flared opening. Here, the wavefront curvature abruptly changes from $K = 0$ to $K = K_{cr} \cdot \sin(\beta)$ and the wave propagates out of the narrow pathway without experiencing a conduction block. When the excitation wave is initiated from the right side of Fig. 3b, it takes on a concave form in the tapered pathway and its curvature abruptly changes from $K = 0$ to $K = K_{cr}$ when the pathway borders become parallel. At the left opening, when $\beta = \pi / 2$, the curvature is $K = K_{cr}$ and a conduction block occurs. When both ends of the narrow pathway have an opening with $\beta = \pi / 2$, waves can enter the narrow pathway from either direction, but will be blocked at the opposite end of the pathway.

When the pathway is tapered over its entire length, the wave propagating from the wider end toward the narrow end experiences a variation of its curvature inside the channel from positive values (see Eq. (8)) to $K = -2 / W$ at the narrow end; if $K = K_{cr} = -2 / W$, then a conduction block may occur. However, when a wave propagates in the opposite direction, the curvature of the wavefront never exceeds the critical value. Hence, the spatial configuration of the pathway is a sufficient determinant for a unidirectional block.

When the excitable myocardium depicted in Fig. 3 has uniformly anisotropic properties it therefore has a lower resistance and more rapid conduction between elements i in the direction parallel to the longitudinal axis of the narrow pathway ($D_x / D_y > 1$). This type of creation of a unidirectional block is more difficult than with isotropic conduction. This is a consequence of the gradient component ($\partial V_m / \partial y$) in anisotropic tissue being smaller than for isotropic tissue; therefore, the wavefront curvature is smaller when it exits a narrow pathway.

It was shown earlier that the wavefront inside a narrow path with boundary conditions $\partial V_m / \partial n = \text{grad}(V_m)$ and with isotropic tissue can be estimated as a semicircle of radius $R = W / 2$ and curvature $K = 2 / W$.

If the width (W) of the narrow pathway is greater than the critical width (W_{cr}), propagation can occur in either direction. If $W \leq W_{cr}$, waves entering from either direction will not be able to exit from the opening in the opposite end of the pathway; this is an example of a bidirectional block. Therefore, a unidirectional block cannot occur in these pathways.

10.3.3. Propagation Through Tapered Pathways

A unidirectional block can occur, however, when the narrow pathway has an appropriate tapered border configuration. Waves propagating from the wide end of the tapered pathway cease to exist when they reach the narrow end, while waves that propagate from the narrow end are able to propagate through unhindered (this phenomenon was observed in computer simulations by A. Pang and B. Billet [10]). In order to explain this phenomenon, let us consider the geometry of a wavefront inside a tapered pathway at points *i-i* and at the narrow end (Fig. 6*A*).

When the boundary conditions are $\partial V_m / \partial n$ = grad (V_m), the vector grad V_m is perpendicular to the boundaries throughout the pathway. The boundaries effectively serve in this case as a tangent to the wavefront.

Assuming as before, that the wavefront inside the narrow tapered pathway can be approximated by a portion of a circle, an expression for wavefront curvature can be obtained:

$$K_i = \frac{1}{(W_n/2)+(L-L_i)\tan(\beta)}\cos(\beta). \tag{11}$$

At the narrow end exit from the pathway, $L = L_i$ and (11) reduces to

$$K_n = \frac{2}{W_n}\cos(\beta). \tag{12}$$

Since it is evident that $K_n > K_i$, it follows that a unidirectional block is most likely to occur in the narrow end of the tapered pathway. If we take into consideration the

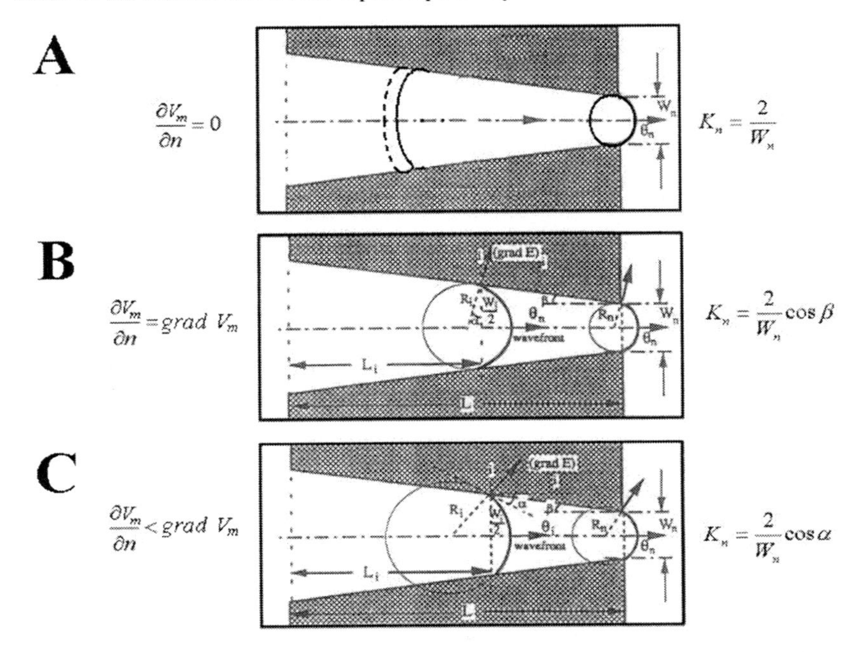

Fig. 6. Wavefront curvatures inside pathways with tapered borders and in narrow exits with three types of boundary conditions.

transient processes of wavefront curvature formation that is associated with abrupt changes in pathway geometry, we can assert that the curvature of the wavefronts crossing the narrow end from left to right (in Fig. 6) is greater than when a wave crosses the narrow opening in the opposite direction. Here, we observe directional differences in propagation due to differences in the length of the wavefront arc relative to the size of the excitable region into which the propagated wave must penetrate. The source-sink concept leads to the same conclusion.

The same reasoning can be applied to the case of narrow pathways with boundary conditions: $0 < \partial V_m / \partial n <$ grad (V_m). A unidirectional block is possible only in a tapered pathway with specific geometric parameters. The expression for the curvature at any border points i inside the tapered path can be obtained by replacing cos (β) in (11) by cos (α_i):

$$K_i = \frac{1}{(W_n/2)+(L-L_i)\tan(\beta)}\cos(\alpha_i)$$
$$= \frac{1}{(W_n/2)+(L-L_i)\tan(\beta)}\frac{1}{\sqrt{1+\tan^2(\alpha_i)}}, \qquad (13)$$

where

$$\tan(\alpha_i) = (\partial E/\partial x)_i / (\partial E/\partial y)_i \qquad (14)$$

At the narrow end exit, $L = L_i$ and the curvature K is equal to K_n:

$$K_n = \frac{2}{W_n}\cos(\alpha_i). \qquad (15)$$

As in the previous case, a unidirectional block can occur when the wave propagates from the wide end toward the narrow end of a tapered pathway. It follows that geometric asymmetry is necessary for a unidirectional block to appear in a narrow pathway when myocardial excitation properties are the same for either direction.

Similar to the case of impermeable borders, certain pathways with border zones exhibiting decreased excitability will cause conduction block when exiting to a larger area of viable media.

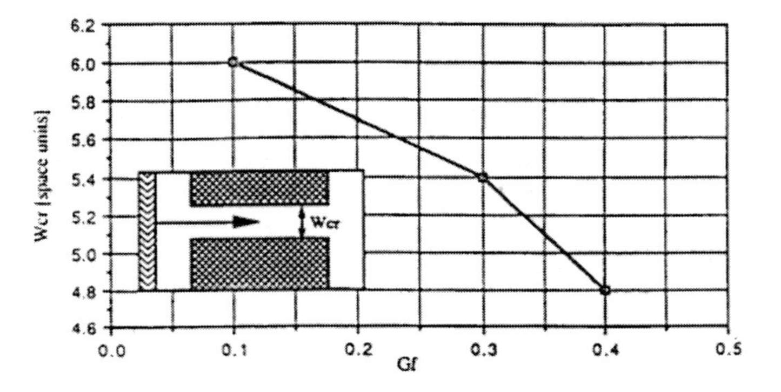

Fig. 7. Dependence of critical width, W_{cr}, on the excitability (G_f) of the narrow pathway borders providing unidirectional block. As G_f increases, the border zone current consumption decreases. Pathway borders are parallel.

Fig. 7 shows computer simulation results where the critical width W_{cr} is determined for various levels of media excitability for the pathway borders.

Computer simulations using the FHN model reveal that for pathways bordered by zones with decreased excitability, it is possible to find combinations of the model parameters and tapered pathway geometric characteristics (G_f, β, and W_n) in which a unidirectional block occurs.

Fig. 8 shows computer simulation data demonstrating that conduction velocity decreases with pathway width.

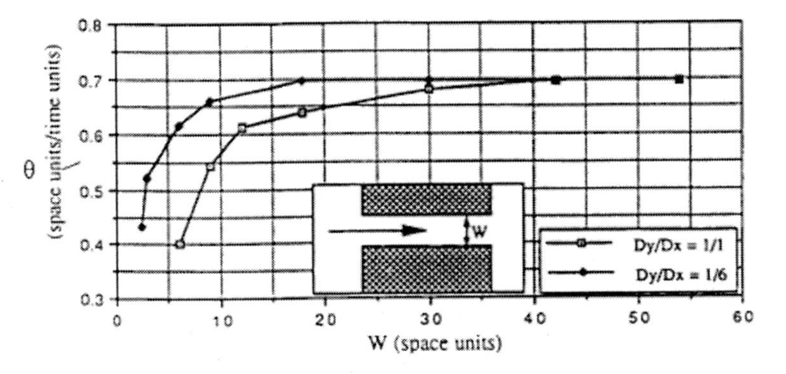

Fig. 8. The relationship of wavefront conduction velocity (θ), inside narrow pathways with border zones of decreased excitability ($G_f = 0.1$), on pathway width (W) is shown for both isotropic and anisotropic ($D_y/D_x = 1/6$) media (measured from computer simulation).

Data for isotropic and uniformly anisotropic ($D_y/D_x = 1/6$) myocardium shows that for pathways of equal width, wave propagation is slower in isotropic media (Fig. 8). Anisotropy also decreases the critical width of the narrow path, which allows for a unidirectional block. This effect of anisotropy can be explained by a decrease in current consumption along the transverse direction when a wave exits from the narrow path into an open viable myocardium. In comparison to the case of isotropic media, this leads to a decrease of wavefront curvature outside the narrow pathway.

The dependence of wavefront conduction velocity on its curvature is an important characteristic of excitable tissue, which cannot be determined by direct experiments. This provided the impetus for the method based on (1) and (2) and considered in detail in [5]. Equations (1) and (2) express the dependence of the curvature K on pathway width W and the ratio of the components of grad (V_m) for narrow pathways with parallel borders and border zones exhibiting decreased excitability. The components of grad (V_m) at the intersection of the wavefront and the pathway border can be determined in computer simulations by calculating the finite difference approximation of the partial derivatives. Fig. 9 shows the results of estimating the curvature for the same pathways as in Fig. 8.

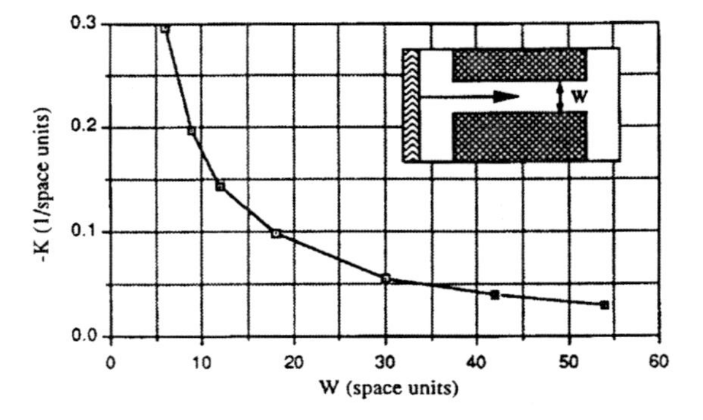

Fig. 9. The relationship of estimated wavefront curvature (K), inside narrow pathways with border zones of decreased excitability ($G_f = 0.1$), on pathway width (W) is shown for computer simulations of isotropic media.

As expected, the curvature decreases as the pathway width is enlarged. The critical curvature K_{cr} is near the maximum value of K shown in this figure. Finally, by combining data from Figs. 8 and 9 and eliminating the parameter W, a relationship between the wavefront conduction velocity and the curvature is obtained (see Fig. 10).

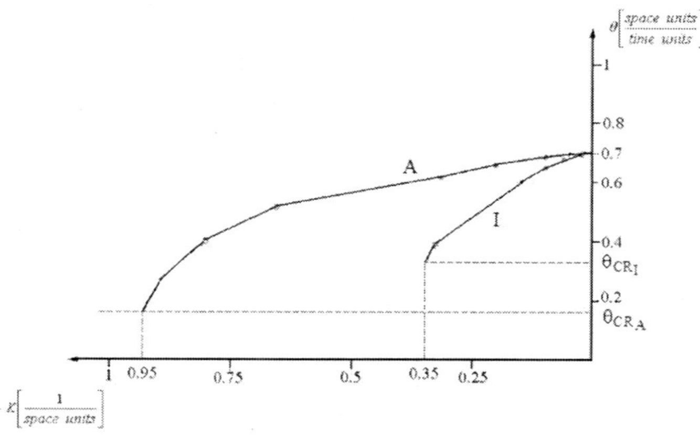

Fig. 10. The relationship of measured wavefront conduction velocity (θ) on estimated curvature (K) for the modified FHN model [11], obtained using data presented in Figs. 8 and 9. In the figure, **A** and **I** denote the anisotropic and isotropic cases, respectively.

The relationship for isotropic media (curve **I**) in Fig. 10 is in good agreement with data obtained for an FHN model with standard parameters [12].

10.3.4. Initiation of Reentry

The appearance of a unidirectional block for excitation wave propagation in cardiac tissue may be initiated by several factors and lead to reentrant arrhythmias [13]. It is possible to achieve such blocks when narrow paths are formed in myocardium by post-infarct scars and the wavefront curvature in at least one of the exits becomes equal to or more than the critical value. In order to obtain reentrant wave propagation, it is necessary to have at least one additional channel which has the ability to conduct waves in both directions. This idea is illustrated in Fig. 12 using the simplest configuration consisting of a narrow pathway with combined parallel and tapered border geometries and impermeable borders.

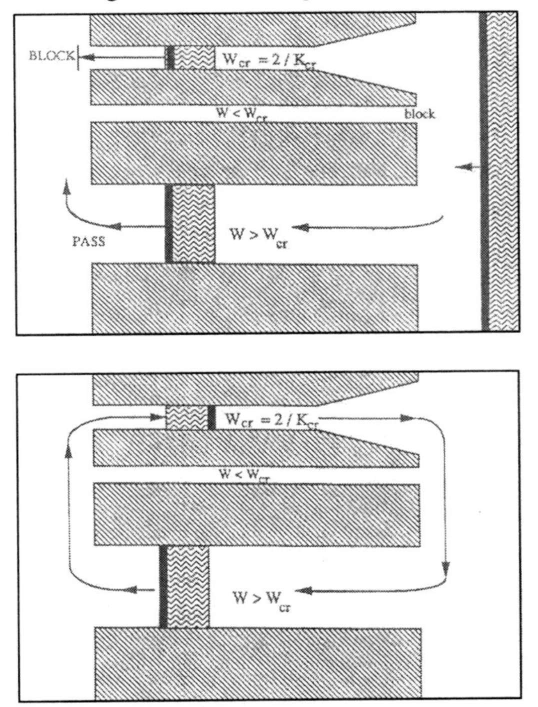

Fig. 12. Effect of geometry in the myocardium on the appearance of a unidirectional block and reentry when the borders are impermeable.

In Fig. 12, two phases of the reentry formation are illustrated: the appearance of a unidirectional block on the upper half of the myocardium and the establishment of reentry after a wave penetrates the lower wide pathway after the recovery processes have ceased in the upper narrow path. In this situation, a proper length for the reentrant path and recovery from inactivation are required for stable reentry to occur. The more detailed explanations for this case and several others are given in comments regarding computer simulation results (see Section 10.4.2).

10.4. Computer Simulation Results

This section covers results of computer simulations concerned with changes in the velocity and shape of rectilinear wavefront excitation through narrow paths with parallel and tapered border geometry both for isotropic and anisotropic viable tissues inside. The effects of viable tissue anisotropy and AP restitution property on reentry appearance are treated using computer simulation. Finally, computer simulation data are presented for the case when narrow path borders have a large current sink property. The effect of this property on reentry is demonstrated.

10.4.1. Transient Phenomena at the Entrance and Exit of Narrow Pathways

The wave front isochrones presented in Figs. 13A and 13B [14] demonstrate that an originally rectilinear wavefront opposite to the narrow pathway entrance becomes curvilinear (with positive curvature) as it moves closer to the pathway opening both for isotropic and anisotropic properties of viable tissue. The difference is that in the case of anisotropy (with the anisotropy ratio $D_x / D_y = 9$), curvature transients are more pronounced (Fig. 13B), but the resultant conduction velocities are identical to isotropic tissue, except that the critical width of the pathway may be much smaller.

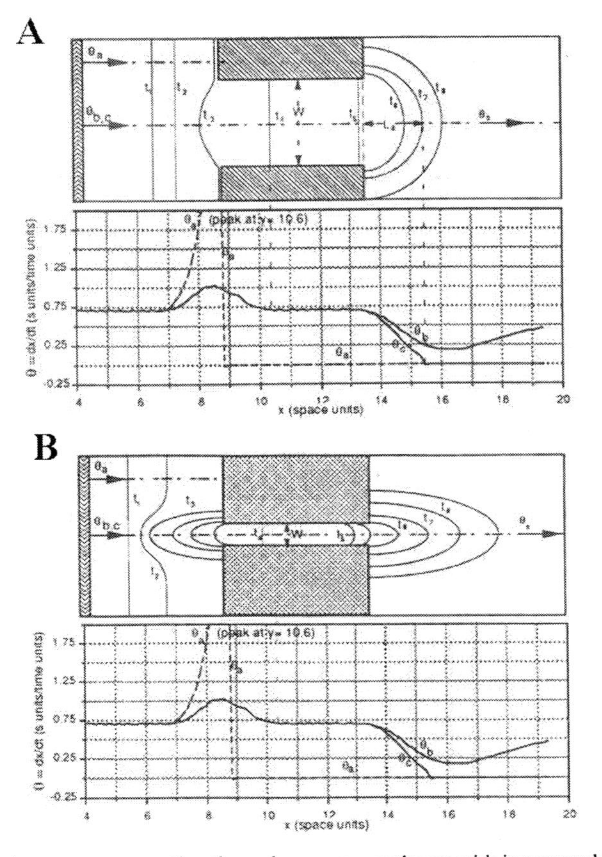

Fig. 13. Excitation wave propagation through a narrow pathway with impermeable parallel borders. θ_a is the conduction velocity of the wavefront measured at a distance of 8 space units from the pathway symmetry axis, θ_b and θ_c are the conduction velocities of the wavefront measured along the pathway symmetry axis, and t_1 –t_8 designate wavefront isochrones. (A) Isotropic tissue, with θ_b given for $W = 2.4$ space units and θ_c given for $W = 2.1$ space units (critical value). (B) Anisotropic tissue ($D_x/D_y = 9$), with θ_b given for $W = 1.05$ space units and θ_c given for $W = 0.9$ space units (critical value).

Wave propagation in tapered pathways with boundary conditions of decreased excitability ($0 < \partial V_m / \partial n <$ grad (V_m)) shows that there exists a width at the narrow end of the pathway such that propagation toward the wide end is possible while it is blocked in the opposite direction. This can be explained by the different characters of wavefront formation processes and corresponding changes in the wavefront curvatures in regions near the narrow end of the tapered pathway for waves crossing it in both directions (see Fig. 14A and B).

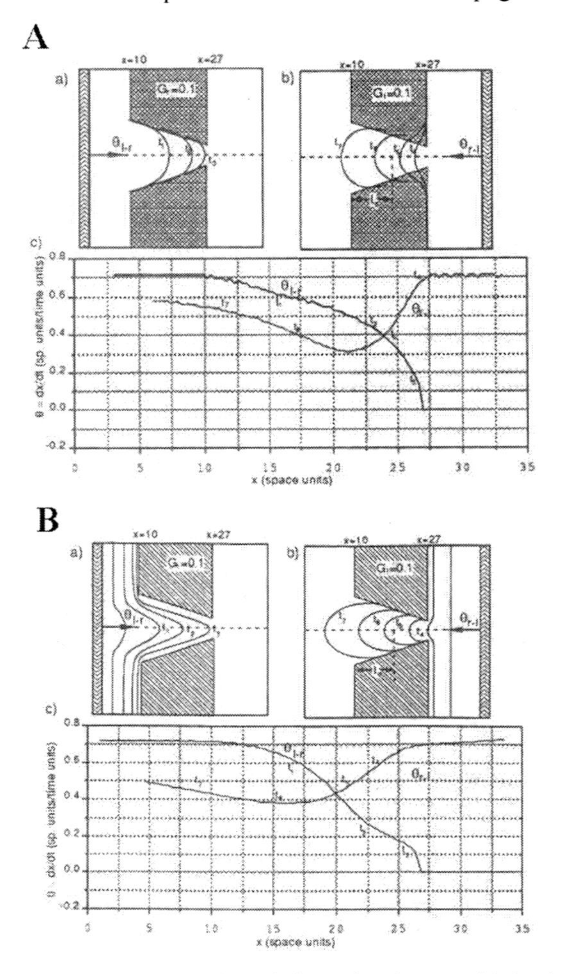

Fig. 14. Excitation wave propagation through tapered pathways with border zones of decreased excitability ($G_f = 0.1$). The wavefront isochrones at t_1 - t_7 are measured at a wavefront potential $V_m = 0.45$ units. $\theta_{i\text{-}r}$ and $\theta_{r\text{-}i}$ are the conduction velocities along the axis of symmetry. (a) Isotropic tissue, $W_n = W_{cr} = 3$ space units. (b) Anisotropic tissue ($D_x / D_y = 9$), $W_n = W_{cr} = 0.9$ space units. (c) Wavefront velocities.

The wavefront morphology measured at the level of $V_m = 0.45$ units at consecutive moments in time are shown in Figs. 14a and 14b for waves moving left to right and vice versa, respectively. The corresponding wavefront conduction velocities were measured along the axis of symmetry in the pathways and are presented in Figs. 14c for both directions of wave propagation. The effect of uniform anisotrop ($D_x/D_y = 9$) is demonstrated in Fig. 14b. As theory predicts, the simulation results show elliptic wavefront morphologies and permit propagation through smaller widths in comparison to cases of isotropic tissue.

10.4.2. Reentrant Propagation in a Myocardium Model with Narrow Pathways

A potential reentry circuit consists of at least two interconnected pathways located between myocardial scars. At least one pathway must provide a unidirectional block, while another must be able to support bidirectional propagation.

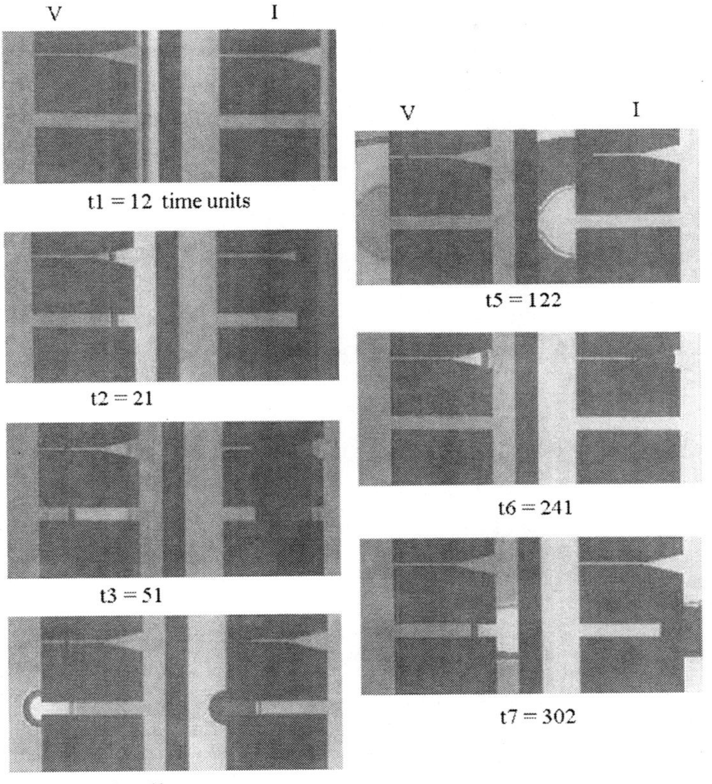

Fig. 15. Time series of reentry in homogeneous, isotropic tissue with fast recovery processes (short APD restitution) and impermeable borders. The tapered pathway length is 48 space units, the width of the narrow portion is 1.8 space units, and $\beta = 30°$. t_1- excitation begins on the right edge, t_2 – the wave enters the narrow parallel and tapered pathways, t_3 - continuation, t_4 - excitation in the tapered pathway is blocked, t_5 - wave reentry occurs at the initial site of block, t_6 – the wave is unblocked as it leaves the tapered pathway, and t_7 – the wave reenters the parallel channel.

Three major cases are studied using a computer simulation: (1) all excitable media is homogeneous, isotropic, and with fast recovery processes; (2) the same isotropic media is utilized, but with slow recovery processes; and (3) all excitable media is anisotropic with slow recovery processes.

This approach allows separate, simultaneous observations of the effects of anisotropy and APD restitution on the appearance of reentry. Fig. 15 shows the sequence of wave propagation (left vertical column) and outward current (right vertical column) for the first case after applying a rectilinear excitation along the right side of the grid. In the absence of long APD restitution, reentry circuits of any size are easily produced.

In the second case (Fig. 16), longer APD restitution is added to the conditions of the first case. Here, it can be seen that the presence of residual outward current (which determines the APD restitution properties) prevents the reentrant wave from penetrating the narrow channel at the initial site of the block and completing the reentry circuit. Thus, the additional delay in APD restitution results in a bidirectional block at the narrow end of the pathway.

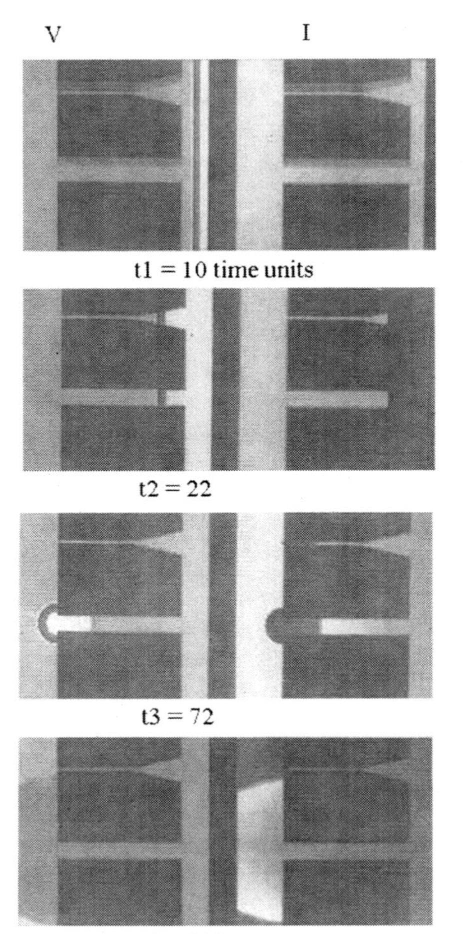

Fig. 16. Time series of reentry in homogeneous, isotropic tissue with slow recovery properties and impermeable borders. t_1 - initial excitation enters from the right edge, t_2 – the wave enters the pathways, t_3 – the tapered pathway blocks propagation, whereas the wide channel is nonblocking, t_4 – the residual current in the tapered pathway blocks the reentrant wave at the initial site of the block.

In the final case with impermeable borders in the pathway, anisotropy (with an anisotropy ratio D_x/D_y = 6) is added to the conditions of the previous case. Anisotropy (Fig. 17), facilitates reentry by introducing an additional time delay for

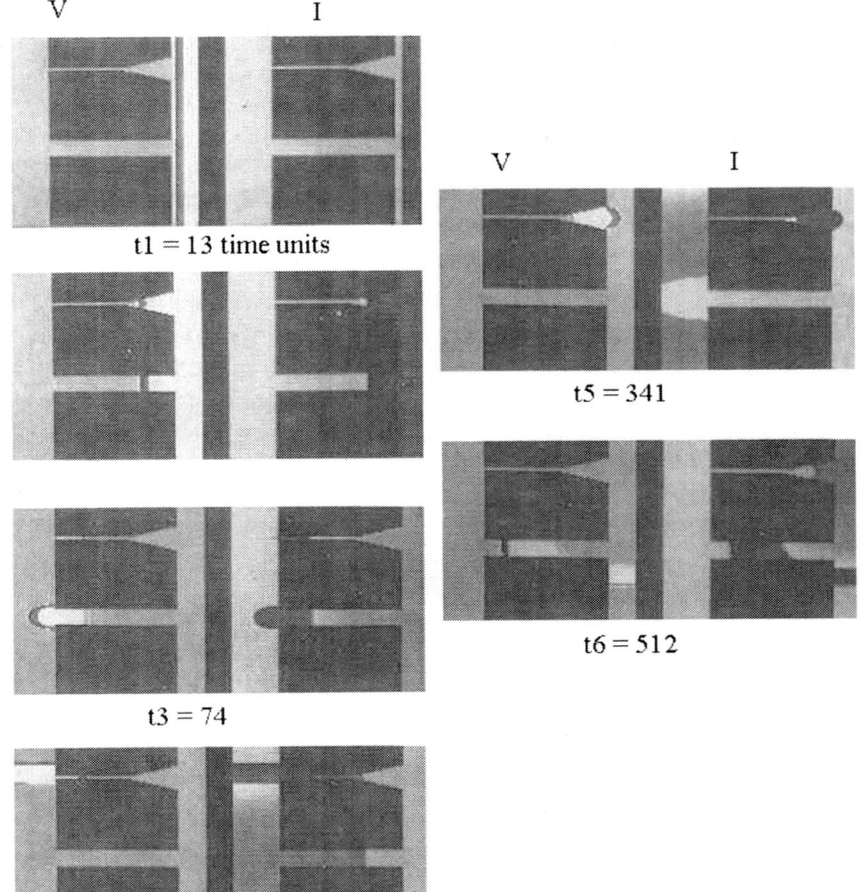

Fig. 17. Time series of reentry in uniform anisotropic tissue with slow recovery properties (ε_4 = 0.022) and impermeable borderes. t_1 - initial excitation enters from right edge, t_2 – the wave enters the pathways, t_3 – the pathway with tapered borders blocks propagation, whereas the wide pathway is nonblocking, t_4 – the residual current in the tapered path is too small to block the reentrant wave, t_5 - reentrant excitation continues through the tapered pathway, t_6 - wave reentry occurs in the wide pathway with parallel borders.

the reentrant wave to propagate in the *y* direction, allowing more time for recovery at the initial site of the block.

The speed of stationary wave propagation in the *y* direction is $\sqrt{6}$ times smaller than in the *x* direction. Thus, the anisotropy of the tissue facilitates reentry in the presence of narrow paths with impermeable borders. Prolonged APD restitution has the opposite effect.

10.4.3. Narrow Pathways with a Large Current Sink

The presence of border zones, which serve as a current sink, can potentially facilitate reentry by slowing conduction or inhibit reentry by promoting a bidirectional block, especially when APD restitution is prolonged. In the simplest case, where the tissue is isotropic and with a shortened APD restitution, reentry can occur even with closely spaced channels (see Fig. 18). When longer APD restitution is introduced into the myocardium model, reentry is prevented due to the presence of a residual outward current from the previous excitation of the tissue. This occurs even when the spacing between the channels is increased (see Fig. 19).

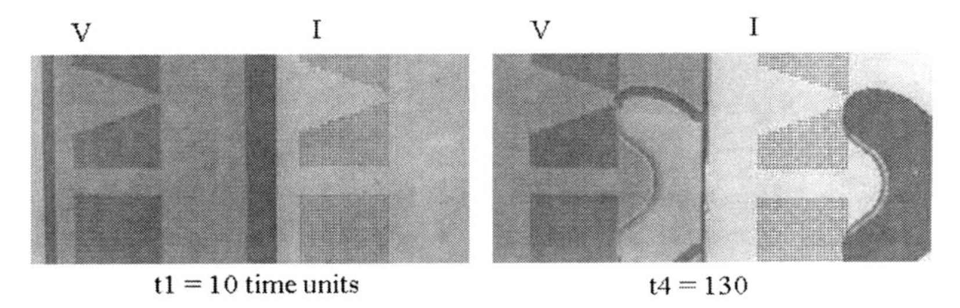

t1 = 10 time units　　t4 = 130

t2 = 45　　t5 = 187

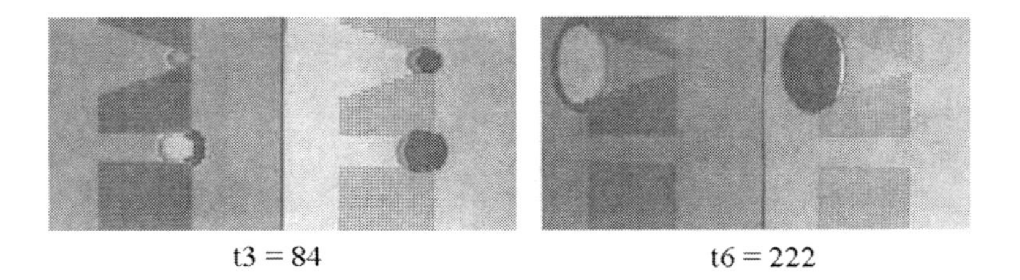

t3 = 84　　t6 = 222

Fig. 18. Similar to Fig. 15, but with shorter APD restitution.

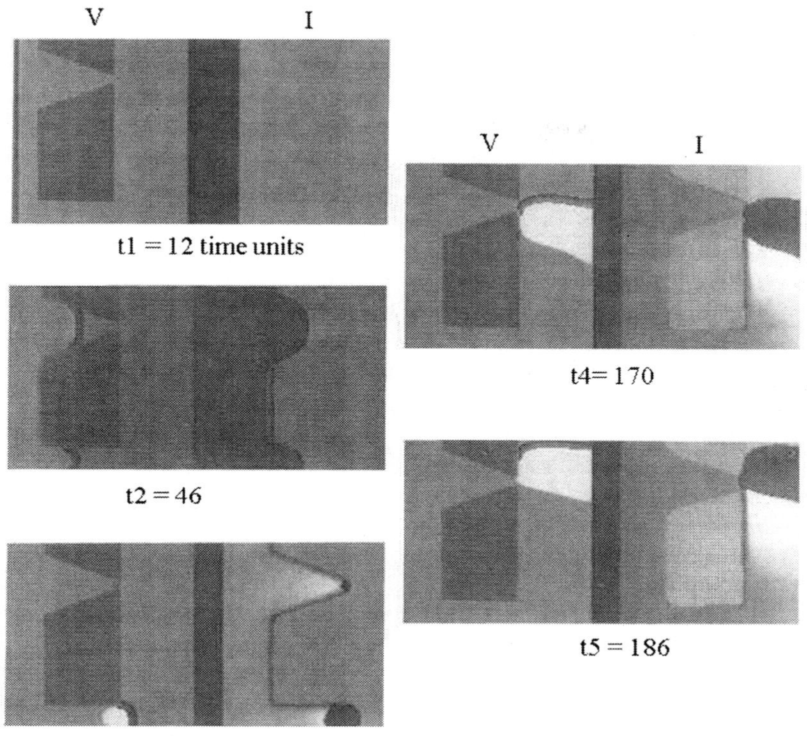

Fig. 19. Time series of failed reentry in media bordered by elements with decreased excitability. Viable media is homogeneous, isotropic, and has prolonged APD restitution (ε_4= 0.022). For border tissue, G_r = 0.1, the path length = 30 space units, β = 20°, and the distance between pathway centers is 52.5 space units. t_1 – initial excitation enters from the left edge, t_2 – the wave enters the pathways, t_3 – the tapered pathway blocks propagation, t_4 – the residual current blocks propagation at the initial site of the block, and t_5 – continuation.

In this simulation, anisotropy did not provide a sufficient delay for avoiding the effects of residual current at the initial site of the block. A greater separation between pathways or the slowing of the conduction would be required to allow reentry. Tissue anisotropy also decreases the grad (V_m) component in the direction transverse to the fiber and increases it in the longitudinal direction. This increases the curvature of the wavefront in the longitudinal direction, making it more difficult for the reentering wave to penetrate the narrow end of the tapered pathway. Therefore, in the case of narrow pathways with "large current sink" border tissue, tissue anisotropy under certain conditions facilitates a bidirectional block and can impede reentry.

10.5. Discussion

The major topic of this chapter revolves around wave propagation through excitable 2D narrow pathways with borders that are impermeable or have decreased excitability. These theoretical considerations are based on knowledge of the wavefront's grad (V_m) components, and the assumption that the wavefront at the exit from a narrow pathway (width ~ W_{cr}) can be approximated as a circular arc. This assumption was verified by computer simulations of the FHN equations, and was shown to be sufficiently accurate. Pertsov *et al.* [16] reached the same conclusion based on computer simulations of waves propagating through holes (with a diameter close to critical) in thin and thick impermeable screens.

Theoretical considerations allow for a connection between the geometry of narrow pathways and their boundary conditions to wavefront curvature. This approach provided the means to estimate wavefront critical curvature.

Computer simulations show that these results are valid for the modified FHN model. Simulations also illuminate the effects of tissue anisotropy and APD restitution on wave propagation in narrow paths, and on the appearance of reentry:

- Reentry is possible for either border condition when a pathway with a unidirectional block exists in parallel with at least one non-blocking pathway.
- APD restitution tends to inhibit reentry. For reentry to occur, an increased time delay for wave propagation in the reentry loop is required.
- For narrow paths with impermeable borders, myocardial anisotropy greatly facilitates the development of reentry (introduces natural delay) and at the same time expands the range of path width (in the direction of smaller width) for which a unidirectional block and reentry are possible.
- For narrow paths bordered by tissue serving as an abnormal current sink, tissue anisotropy can facilitate a unidirectional block by increasing the arrival time of the excitation wave front at the initial site of the block, or it can inhibit formation of a unidirectional block by promoting conduction out from the narrow path.

10.6. Conclusion

The theoretical considerations and computer simulations results presented in this chapter demonstrate that:

a. Specific geometric configurations of narrow pathways exist in 2D tissue, which facilitate the appearance of a unidirectional block and reentry of excitation. This remains true even when membrane properties of the viable tissue are normal.
b. The effects of pathway geometry are strongly dependant on the boundary conditions (between scars and viable tissue).
c. Computer simulations of excitation propagation in narrow paths permit a determination of the curvature relation ($\theta = f(K)$).
d. Recovery processes inhibit the reentry of excitation regardless of the type of boundary conditions.
e. The uniform anisotropy of viable tissue can facilitate reentry in cases with impermeable borders.

The revision of the results for viable myocardium that incorporates an up-to-date ionic AP model with developed intracellular Ca dynamics represents a topic of significant theoretical and practical interest. More over the local reentrant circulation in the presence of myocardial scars may cause ventricular fibrillation without premature excitation.

10.7. References

1. de Bakker, J.M., F.J.L. van Capelle, M.J. Janse, A.A. Wilde, R. Coronel, A.E. Becker, K.P. Dingemans, N.M. van Hemel, and R.N. Hauer, *Reentry as a cause of ventricular tachycardia in patients with chronic ischemic heart disease: electrophysiologic and anatomic correlation.* Circulation, 1988. **77**: 589-606.
2. Bolick, D.R., D.B. Hackel, K.A. Reimer, and R.E. Ideker, *Quantitative analysis of myocardial infarct structure in patients with ventricular tachycardia.* Circulation, 1986. **74**: 1266-1279.
3. Fenoglio Jr., J.J., T.D. Pham, A.H. Harken, L.N. Horowitz, M.E. Josephson, and A.L. Wit, *Recurrent sustained ventricular tachycardia: structure and ultrastructure of subendocardial regions in which tachycardia originates.* Circulation, 1983. **68**: 518-533.
4. de Bakker, J.M., R. Coronel, S. Tasseron, A.A. Wilde, T. Opthof, M.J. Janse, F.J. van Capelle, A.E. Becker, and G. Jambroes, *Ventricular tachycardia in the infarcted, Langendorff-perfused human heart: role of the arrangement of surviving cardiac fibers.* J Am Coll Cardiol, 1990. **15**: 1594-1607.
5. Kogan, B.Y., W.J. Karplus, B.S. Billet, and W.G. Stevenson, *Excitation wave propagation within narrow pathways: geometric configurations facilitating unidirectional block and reentry.* Physica D, 1992. **59**: 275-296.
6. Gilmour Jr., R.F., J.J. Heger, E.N. Prystowsky, and D.P. Zipes, *Cellular electrophysiologic abnormalities of diseased human ventricular myocardium.* Am J Cardiol, 1983. **51**: 137-144.
7. Myerburg, R.J., K. Epstein, M.S. Gaide, S.S. Wong, A. Castellanos, H. Gelband, J.S. Cameron, and A.L. Bassett, *Cellular electrophysiology in acute and healed experimental myocardial infarction.* Ann N Y Acad Sci, 1982. **382**: 90-115.
8. Dillon, S.M., M.A. Allessie, P.C. Ursell, and A.L. Wit, *Influences of anisotropic tissue structure on reentrant circuits in the epicardial border zone of subacute canine infarcts.* Circ Res, 1988. **63**: 182-206.
9. Zykov, V.S., *Analytic estimate of the dependence of excitation wave velocity in a two dimensional excitable medium on the curvature of its front.* Biofizika (USSR), 1980. **25**: 888-892.
10. Pang, A.T., *On Simulating and Visualizing Nonlinear Distributed Parameter Systems Using Massively Parallel Computers*, in *Computer Science*. 1990, University of California, Los Angeles: Los Angeles. p. 155.
11. Kogan, B.Y., W.J. Karplus, B.S. Billet, A.T. Pang, H.S. Karagueuzian, and S.S. Khan, *The simplified Fitzhugh-Nagumo model with action potential duration restitution: effects on 2D-wave propagation.* Physica D, 1991. **50**: 327-340.
12. Zykov, V.S., *Simulation of Wave Process in Excitable Media.* Nonlinear science: theory and applications, ed. A.V. Holden. 1987, Manchester and New York: Manchester University Press.
13. Cabo, C. and R.C. Barr, *Unidirectional block in a computer model of partially coupled segments of cardiac Purkinje tissue.* Ann Biomed Eng, 1993. **21**: 633-644.
14. Kogan, B.Y., W.J. Karplus, and M.G. Karpoukhin. *The effect of boundary conditions and geometry of 2D excitable media on properties of wave propagation.* in *International Workshop on Dynamism and Regulation in Non-linear Chemical Systems.* 1994. Tsukuba, Japan: National Institute of Materials and Chemical Research (Japan). p. 79-81.
15. Kogan, B.Y., W.J. Karplus, and B.S. Billet. *Excitation wave propagation through narrow pathways.* in *Spatio-Temporal Organization in Nonequilibrium Systems.* 1992. Berlin, Germany: Project Verlag. p. 122-127.
16. Pertsov, A.M., E.A. Ermakova, and E.E. Shnol, *On the diffraction of autowaves.* Physica D, 1990. **44**: 178-190.

Concluding Remarks

This book has treated the fundamental problems of mathematical modeling of electrophysiological processes responsible for generation of AP in cardiomyocytes and their propagation through myocardium. These processes, together with cells' ability to contract, are responsible for providing the primary heart function of pumping blood through whole organism.

From a structural point of view, the myocardium represents a system of discrete excitable and contractile elements, myocytes. Gap junctions provide one of the key interconnects between these elements. The size of a myocyte is on the order 10 nanometers while the scale of cardiac tissue is incomparably bigger. Thus the average cell properties of AP generation, concentrated at each point of the myocardium and connected through intra- and extra-cellular liquid resistance are usually assumed for investigation the wave processes in myocardium. The spatially distributed intra-cellular properties are typically neglected. Under normal conditions, the resistance of a gap junction is much smaller than the resistance of intra- cellular liquid and it is possible to consider all cardiac tissue as syncytium, a continuous system, where AP propagates according to diffusion properties.

Mathematical modeling of these systems is reduced to the solution of a special type of nonlinear reaction-diffusion equations. Obtaining these solutions in analytical form is very difficult, even for simplified cases, and is impossible for more realistic cases. Moreover, some relationships are not known and are introduced into models as analytical expressions obtained by fitting to results of particular physiological experiments (semi-phenomenological models).

Therefore computer simulations are required to obtain qualitative and quantitative results. Due to the enormous computational complexity, massively parallel supercomputers are needed, even today, for most 2D and all 3D problems. Special numerical algorithms, which include the application of adaptive time and space steps for a given grid representation of tissue, are required to obtain computationally tractable results.

When stimulus is applied to the extracellular domain of tissue (e.g. in a case of defibrillation), it is necessary to introduce the bi-domain tissue representation. Except for the case of fully uniform tissue, this representation requires an additional, simultaneously, solution of an elliptic PDE for the same tissue grid. Experience shows that an advanced multigrid sequential algorithm approximately doubles computation time (R. Samade personal communication). Thus development of an efficient version of the multigrid algorithm suitable for parallel execution remains an outstanding problem.

Mathematical modeling and computer simulation of complex problems are a powerful method for scientific investigation not only in physics, engineering, but in biology and medicine as well. In order to obtain successful results the subject and major goals of investigation must be precisely formulated. At the same time all assumptions and restrictions must be mentioned, including the conditions under which the experimental data used for the simulation were obtained. A good mathematical model can predict new phenomena but this does not mean that the results of simulation can be extrapolated to the cases not covered by the used model.

For example, it is impossible to judge about results of Ca dynamics using models where it is not represented or is represented in rudimentary form.

Validation of the modeling results is one of the important subjects. Even today, direct comparison with physiological experiments is difficult if not impossible for some cases because the shape and heterogeneity of a real heart are very different from that using in simulation. For example, small blood vessels anchor the spiral waves; cells change their directions and so on. Physiological experiments, especially with a tissue, cannot measure the all internal variables of a mathematical model and accuracy of measured values is not high enough. It is worthwhile to remember that almost all mathematical models were developed to reproduce the heart functions under normal conditions when all the processes are close to stationary or quasi-stationary. These relate to obtaining the gated channel currents using constant clamp voltages and use of stationary expression for C_aIC_aR processes from JSR (LRd and Chudin models). New phenomena appeared during tachycardia and fibrillation connected with Ca accumulation in SR and sarcoplasma, which lead to appearance of EADs and DADs clusters on the pattern of cells, occurring under high pacing rate. The latter also lead to changing the character of C_aIC_aR process in SR from static to dynamic and under some conditions may cause Ca and AP alternance (see [22] in chapter 4). Thus many questions are left unanswered about the correctness of applying an AP model developed for normal conditions to pathological cases. Here appropriately to compare at least the results obtain for membrane channel gates controlled by AP under different pacing rates. Very little information is available about mechanisms of Ca release from SR, especially, about spontaneous release caused by overloading of SR and intracellular domain with Ca.

Results from physiological experiments for AP and wave propagation under normal conditions were used to validate the mathematical model. Instead of discarding and disregarding new phenomena observed using the model under abnormal conditions (see chapter 4 and [28] where spontaneous release is not presented) it is worthwhile to create a plausible hypothesis about the unknown mechanism and continue the investigation further. As new experimental data becomes available, the hypothesis may have to be reconsidered. It may be recognized to be incorrect or to have restricted application. This approach is widely used in physics and other branches of sciences connected with real world.

Application of mathematical models and computer simulations produced many fruitful predictions. Spiral waves, discovered in the author's lab (see [3] in chapter 8), in 2D cardiac tissue, were later observed in real tissue. Simulations results found and explained the appearance of EAD and DAD clusters in 2D tissue during spiral wave propagation, proved that EAD and DAD can appear in single cell not only in case of long but also under short period of stimulation.

I hope that publication of this book will attract new researchers to the application of mathematical modeling and computer simulation of biological system and, in particular, generate more attention to cardiology problems.

Exercises

1. Prove that linear second order oscillator

$$\frac{d^2u}{dt^2} + 2\alpha\frac{du}{dt} + \omega_0^2 u = 0$$

with initial conditions: $u(0) = 7$, $\dot{u}(0) = 0$ and parameters $\alpha = -1$, $\omega = 10$ produces oscillations with amplitude increasing over time. Explain why the amplitude of oscillations will be finite in real systems.

2. Ionic currents through a membrane can be expressed using two formulations:
    a. the Hodgkin-Huxley (HH) formulation
    b. and the Goldman-Hodgkin-Katz (GHK) formulation.

What formulation would you choose if the extracellular and intracellular concentration of the considered ions are variable?
Show that both current formulations give the same results when $V_m = V_{m,S}$.

3. Calculate the rest potential using the GHK equations if Na^+, K^+, and Cl^- ions participate in the ionic currents. Use the data about ionic concentrations and ion channel permeability ratios given in the previous chapter.

4. Derive the relationship between I_{st} and T_{st} for a given $V_{m,th}$. Explain the restrictions applied to the values of T_{st}.

5. Given the definition of the length constant λ. Write the relationship between the length constant and the cardiac cell's parameters. Calculate the length constant for local propagation in a 1D fiber if the cardiac cell is considered to be a cylinder with a radius $a = 8$ μm, $R_i = 200$ Ω-cm, and $R_m = 6.25$ kΩ-cm^2.

6. Calculate the diffusion coefficient using the values of λ and R_m from the previous question. Assume that $C_m = 1$ μF/cm^2.

7. To what category of mathematical models is it possible to relate action potential (AP) models? Explain your reasoning.

8. Find the value of the length constant, λ, for passive propagation in tissue with the following parameters. (See chapter 7, page 128)
$a = 8 - 10\mu M$
$R_i = 0.2$ [kΩ cm]
$R_m = 6.25$ [kΩ cm^2]

9. Which case is excitation conduction velocity larger: for the monodomain or bidomain approach?

10. FitzHugh and Nagumo derived their simplified model from the van der Pol equation.
    a. Show on a phase-plane plot (w, v) the changes introduced by FitzHugh in order to reproduce the nerve AP.
    b. Investigate the stability of the steady state point on the phase-plane plot.

11. The computational solution of mathematical models requires the selection of effective numerical algorithms, adequate computer architectures, and programming tools for visualizing data and calculating complex inherent characteristics of the model, such as the conduction velocity, relaxation coefficient, the AP duration restitution curve, etc.
    a. Explain the rationale for using the operator splitting algorithm in the numerical solution of parabolic partial differential equations.
    b. Describe the hybrid method and compare it with the Euler method for solving ordinary differential equations.
    c. Explain the computer simulation approach for calculating conduction velocity on a given point of a two-dimensional (2D) wavefront.
    d. Using the Noble AP model for propagation in 2D myocardium, show that it is possible to reduce this model into dimensionless form.

12. Stationary propagation of a spiral wave is considered to correspond to ventricular tachycardia, the precursor to ventricular fibrillation.
    a. Given the definitions of the points q and Q on the front of wave exhibiting stationary propagation.
    b. Define the area that is termed the *core* of a spiral wave. What geometric form does the core of the spiral wave have?
    c. Explain why stationary propagation of a spiral wave is impossible in a square-shaped section of myocardium with a restricted size.

13. Excitation-propagation through narrow passes are frequently observed in myocardium following an episode of myocardial infarction, when surviving regions of tissue are surrounded by damaged or dead regions.
    a. What are the three types of boundary conditions characteristic of narrow passes? Give their mathematical formulations and physical meanings.
    b. What geometrical configurations in narrow passes facilitate the appearance of reentrant propagation? What are the necessary conditions for reentry?
    c. How does wavefront curvature inside a narrow path and at its openings depend on the geometry, boundary conditions, and properties of surviving regions of myocardium.

14. According to Courtemanche and associates, the membrane capacitance for an atrial cell with dimensions $L = 100$ μm and $D = 16$ μm is equal to $C_m = 100$ pF. What is the specific capacitance, expressed in μF/cm^2, corresponding to this value of C_m? Note that 1 μm = 10^{-4} cm and 1 pF = 10^{-6} μF.

Index

Breinigsville, PA USA
30 December 2009

229708BV00004B/13/P